London
A New Metropolitan Geography

London

A New Metropolitan Geography

Edited by
Keith Hoggart and David Green

Edward Arnold
A division of Hodder & Stoughton
LONDON NEW YORK MELBOURNE AUCKLAND

© 1991 Keith Hoggart and David R. Green

First published in Great Britain 1991

Distributed in the USA by Routledge, Chapman and Hall, Inc.
29 West 35th Street, New York, NY 10001

British Library Cataloguing in Publication Data

London: a new metropolitan geography.
  1. London. Geographical features
  I. Hoggart, Keith   II. Green, David R.
  914.21
  ISBN 0-340-49319-4

Typeset in Sabon 10/11 pt by Anneset, Weston-super-Mare, Avon
Printed and bound in Great Britain for Edward Arnold, a division of
Hodder and Stoughton Limited, Mill Road, Dunton Green, Sevenoaks,
Kent TN13 2YA by Biddles Limited, Guildford and King's Lynn

Contents

Preface

London today, no less than in the past, generates confusion in the minds of those who live and work in it, as well as those who study it. Recent changes in economy, politics and society have added to this complexity. Like a kaleidoscope in which one move alters a whole pattern, processes of change in London are intricately inter-woven with one another to reveal a fascinating spectrum of inter-related, but distinctive, issues for academic investigation. It is in the hope of unravelling some of the complexities of life in Britain's capital that this book has been written. Yet the intention in presenting the essays in this text is not to offer an encyclopaedic view on every aspect of London life. Instead the book concentrates on key issues in the economy, polity and society of London, with a view to understanding the nature and the implications of areal differences within the city. Chapters have been written with a view to providing analysis and discussion of London life in a manner that offers detail but still provides a commentary that is accessible to the general reader and to students of advanced urban studies courses. This volume is not a textbook, but a reference guide for those wishing to deepen their understanding of the changing character of London.

But what is London? We have sought to maintain certain conventions over the use of this word. 'London' *per se* refers throughout to the area that was covered by the Greater London Council until its abolition in 1986 (viz. to the 32 London boroughs plus the City of London). The 'city' and the 'capital' likewise refer to the Greater London Council area. Figure 11.1 shows the areas that this designation covers, as well as identifying the local government areas that have existed in this area since the Second World War. In contrast, the 'City of London' refers to the square mile that comprises the traditional heart of the capital and today exists as an independent local government unit (although its population in 1981 was only 5864). Overlapping with this geographical area, although by no means coincident with it, is the set of financial institutions that is often referred to as 'the City'. This convention has been retained here, so that the word 'City', with a capital 'C', refers to a set of institutions, rather than a geographical area or local government unit. Two other designations of London have been employed regularly in this book. The first of these is 'inner London', which refers to the 12 London boroughs (plus the City of London) which together cover the area administered by the London County Council until 1965, when it was disbanded. This area is also the same as that administered by the Inner London Education Authority (ILEA) from 1965 until its abolition in 1990. 'Outer London' comprises the 20 boroughs that, with inner London, made up the area that used to be covered by the Greater London Council.

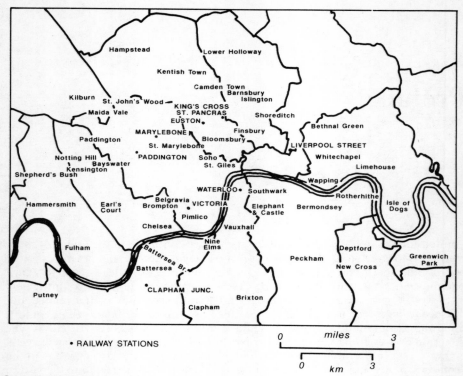

Central London districts

On account of the peculiarities of data availability, in some chapters minor adjustments have had to be made to these definitions. Where this is the case, explicit reference to the changes has been made, and the changes themselves have been described. To help the reader further the figure above shows some of the main districts in the central area of the metropolis.

In preparing this volume we have run up a fair share of debts of gratitude. Susan Sampford started this book on its publication run as geography editor at Edward Arnold. She has our thanks for the patience, friendship and assistance she offered until the move of Edward Arnold's offices to the outer metropolitan belt brought a change in editorial personnel. Her place was taken by Christopher Wheeler, and we thank him for his forebearance in the final stages of the preparation of the manuscript. Our cartographic colleagues at King's College London and the London School of Economics have worked with great diligence and patience in producing the diagrams for this text – and merit special thanks. Finally, we should thank the students of the KCL-LSE Joint School of Geography, for they have had to put up with the distractions of their lecturers as this book was in preparation, and no doubt will find that they are expected to analyze, criticize and extend its contents.

David Green
Keith Hoggart

Figures

Tables

Abbreviations

CIPFA	Chartered Institute of Public Finance and Accountancy
DOE	Department of the Environment
EC	European Community
GATT	General Agreement of Tariffs and Trade
GDP	Gross Domestic Product
GLC	Greater London Council
GLDP	Greater London Development Plan
HMSO	Her Majesty's Stationery Office
ILEA	Inner London Education Authority
LB	London Borough
LCC	London County Council
LPAC	London Planning Advisory Committee
LRT	London Regional Transport
LTE	London Transport Executive
OECD	Organization for Economic Cooperation and Development
OPCS	Office of Population Censuses and Surveys
PPG	Planning Policy Guidance
RB	Royal Borough
SERPLAN	London and South East Regional Planning Conference

Contributors

Bennett, Robert J.	Professor, Department of Geography, London School of Economics
Diamond, Derek R.	Professor, Department of Geography, London School of Economics
Duncan, Simon S.	Lecturer, Department of Geography, London School of Economics, and Visiting Research Fellow, Centre for Urban and Regional Studies, University of Sussex
Frost, Martin E.	Lecturer, Department of Geography, King's College London
Green, David R.	Lecturer, Department of Geography, King's College London
Hamilton, F. E. Ian	Senior Lecturer in East European Studies, Department of Geography, London School of Economics and School of Slavonic and East European Studies
Hebbert, Michael	Senior Lecturer in Planning Studies, Department of Geography, London School of Economics
Jones, Emrys	Emeritus Professor, Department of Geography, London School of Economics
Hoggart, Keith	Lecturer, Department of Geography, King's College London
Morgan, Barrie S.	Senior Lecturer, Department of Geography, and Director, International Students' Office, King's College London
Pharoah, Tim	Lecturer, Department of Town Planning, South Bank Polytechnic
Warnes, Anthony M.	Reader, Department of Geography and Age Concern Institute of Gerontology, King's College London

1

London as an Object of Study

Keith Hoggart

Why publish another book on London? Forests have already been decimated in the name of informing the world about this city. It is not simply what London is, but also what it has been and what it might be that seemingly has captivated writers, publishers and the book-buying public. Editions have emerged on London's lost rivers (Barton, 1962), its 'gentlemen's' clubs (Lejeune, 1979), the open spaces (Forsham and Bergstrom, 1986), its historic hotels (Arnold, 1986), the subterranean world (Trench and Hillman, 1984), the luxury shops (Courtney, 1987), scenes of murders and other crimes (Quennell, 1983; Lane, 1985), even its sign boards (Head, 1957). Want to know about the London of John Betjeman or instead, of Gilbert and Sullivan (Denton, 1988; Goodman, 1988)? How about the city's architecture (Nairn, 1988; Saunders, 1988)? Property might be more in your line (Essberger, 1987; Segrave, 1989)? Failing that, book shops can offer enticements for serious shoppers (Reiber, 1988), for those intrigued by the history of postcards (Finlay, 1986), for reminiscences about war (Mack and Humphries, 1985), or, failing these, books on ethnic London (Zeff, 1986; McAuley, 1987). Realistically, with all this choice, why add to it? The answer lies in the absence of an analytical intention behind these texts. These volumes are designed to occupy space on the coffee tables of Hampstead professionals, Docklands yuppies and Bromley clerks. They are intended to be leisurely reading rather than evaluative commentaries on the causes and consequences of stability and change within the metropolis. Of course, the fact that they exist provides a clear message on the question which opens so many academic treatises on this city; namely, what is London? The variety of content, style and purpose of these coffee table volumes symbolizes the essential message that London *per se* does not exist. 'London' is a concept, the visual, definitional, normative and experiential elements of which rest in the eye of the beholder.

Yet amidst the diversity of conceptualizations that exist there is a strong central element. Embedded within most ideas on what London is, there is an appreciation that London symbolizes much more than the physical fabric and lifestyles of an urban conurbation in southeastern England. Images of London capture messages about the heart of Britain; the *idea* of London is central to the self-image of British people, just as it deeply penetrates the rest of the world's views on British life. As any traveller knows, in bars, restaurants and trains throughout the world, discussion and questions about Britain are infused with messages on the countryside, the weather and royalty. London stands alongside these, and in a sense even surpasses them, for this single word undoubtedly encapsulates what detailed knowledge most foreigners have about Britain and ultimately what induces many to become tourists there. To understand London is to grasp the significance of its central position in the economy, polity and society of Britain. Yet at the same time, what must inform this understanding is London's diversity and turbulence. Each of these can only be understood in the context of the other, for London's national centrality and its compositional differentiation promote forces that are at times mutually reinforcing, at times antagonistic, but always interdependent (e.g. Chapter 10). An understanding of London provides a firm basis for knowing about British society as a whole. But, then, anyone who claims to understand London must have just begun to scratch at the surface. In this text more than the surface covering is analyzed, but in doing so both the certainties and the uncertainties are necessarily brought into focus. London is not so much an object of study as a challenge for investigators (Chapter 13).

London as 'the Centre'

Since the twelfth century at least, London has been the premier city of England (and 'after unification' of Britain as a whole). Its dominance is visualized in the eyes of the general public by its standing as the largest city in the land, as the seat of government, the principal residence of the royal family, the heartland of the British economy and the social and cultural centre of the nation. Associated with these evaluations there is often a normative overtone. London is accused of receiving undue benefits from government. Arts Council awards are concentrated there. Motorways and rail services radiate from it. Investment in its airports seem to expand exponentially, with the North decrying this centralization and the neglected potential of Birmingham, Manchester and Glasgow. Private corporations, seemingly marked by sloth when asked to support non-profit institutions, find that pens glide to cheque books more readily if the 'charity' is a London based symbol of 'culture' (like the opera or the symphonies); provincial institutions charge that they could live off London's crumbs, if only the more wealthy would throw their morsels more than ten miles from Trafalgar Square. In sport, London also takes the lion's share of prestige events. Wimbledon is there, effectively all England's international rugby union and soccer games are played there, this is the only city with two test match grounds for cricket and key symbols of an upper-crust lifestyle – Ascot, Henley, the boat race – are there or just round the corner. One cannot conceive of royalty being married at Birmingham, of the Chancellor making important annual statements to the chamber of commerce in Bristol or of major concert appearances being consistently restricted to Leicester. Diversity in the location of premier events is a hallmark of some nations (like Italy, the USA and West Germany), but, as with Paris, London is too dominant within its nation's boundaries for such diffuseness to be 'tolerated'.

To study London then is to investigate a city with a special place in British society. Appreciating what brought this into being holds one key to understanding the present roles and the persisting distinctiveness of Britain's capital. Yet to account for London's peculiar standing requires attention to events which stretch back at least until the eleventh century. It also involves acknowledging the unusual standing of England itself, for compared with other countries England solved crucial questions of nation-building very early. Central authority, which was evident under Anglo-Saxon monarchs in the tenth century, was confirmed after the Norman conquest of 1066. The establishment of the monarchy and the government in London from that time proved critical. Not unrelated, but significantly adding to the primacy of the city, was the importance of its port. Back to the Middle Ages, around the monarchy an aura developed which set London apart from other centres. For the large landowners, merchants and (later) industrialists, no matter where 'home' was, London was the real centre of life. Although the state apparatus was not highly developed, since the custom of the Crown was to bestow favours as monopoly control over markets, London was a crucial arena for merchantry. Probably more important, as readers of Jane Austen's novels are only too familiar, it was the London 'season' which raised the heart beat of the aspired and the aspiring. This is where real social standing was won or lost. Yet the social elite which lemming-like headed toward London each year was equally important as an economic market (see Chapter 2). Here the port of London was essential, for its proximity and facilities encouraged the expansion of overseas trade and merchantry. Such enterprises were ennobled by aristocratic associations. But trade required negotiable currency, insurance, bridging loans and binding contractual obligations. Thus, London's social scene was a catalyst for the growth of merchant banking, insurance and commodity brokerage with an explicit overseas orientation (i.e. 'the City'). Central to these innovations and expansions were aristocratic connections. Consequent upon these developments was the foundation for the distinctiveness of the economic, political and social environments of London, and indeed in good measure of the whole of Britain.

One of the peculiarities of London was that during the industrial revolution, when Britain led the world in manufacturing performance and dominated international trade, manufacturing within the city was of a small, disorganized character (Chapter 2). In a sense, despite having infrastructure and labour supplies conducive to industrialization, the industial revolution passed London by, at least in its manufacturing form. However, as Geoffrey Ingham (1984) has shown through detailed analysis, on the commercial front the industrial revolution was critical to London's future. The massive expansion in manufacturing trade magnified demand for the City's commercial and financial expertise. As Britain's share of world trade declined, the City was largely unaffected, for no other nation (or city) had developed institutional frameworks of a sufficiently coherent kind to challenge London's position. Indeed, with the City developing strong links with the British government (initially through the negotiations and loan awards which were sought to cover Britain's huge national debt), commercial and financial interests came to influence economic affairs and foreign policy more forcefully; even to the detriment of industrial concerns. The City of London became the banking and arbitration centre around which the world's international trade was oriented. Only in the twentieth century has London's position really been challenged in this regard (by New York) and even this threat has so far been staved off by the willingness of British governments to place their international role above national interests and by the emergence of the Eurodollar market consequent upon enormous US trade deficits (Chapter 5). In this critical area, there is a sense in which London has operated outside the mainstream of the British economy rather than having a complementary role to other geographical areas (King, 1990).

Yet this picture is not a complete one, for in the twentieth century London has come to take on a more central position even in manufacturing. The dominance of commercial and financial interests in Britain has partly resulted from the characteristically weak organization of industrial capital in the nation (Lash and Urry, 1987). With economic difficulties mounting in the 1920s and 1930s, a new impetus emerged in Britain which led to the merging of companies to create larger, more competitive units, with the state lending a hand through new national investments to increase efficiency (e.g. the formation of the Central Electricity Board in 1926). The significance of nationalist (including governmental) imperatives at this time, aligned with the growing involvement of the City in orchestrating and enabling corporate growth, meant that London occupied a forward position in the emergent corporation-dominated national economy (see Chapter 4). Where corporate concerns controlled enterprises through the length of the land, and commonly in other countries as well, the advantages of access to government and finance in London became obvious. Added to which, in the 1920s and 1930s, new consumer-based industries were emerging in Britain on a larger scale than in the past. These were not tied to natural resource distributions, nor to the old industrial heartlands, but sought positions proximate to their main markets. In this regard London was highly placed (Chapter 2). Indeed, in the 1930s the capital and its surrounding territories appeared as an island of growth amidst a general sea of industrial gloom. Economic upswings in the 1950s and 1960s changed this picture somewhat, but, most especially in the 1980s, this pattern is again making an appearance. Here, a critical stimulus to the development of London-centred high technology industries has been access to government defence contracts, alongside the advantages in superior public facilities which proximity to London brings (Hall *et al.*, 1987). For at least the last 60 years, access to finance capital and to the government has induced new industrial endeavours to bloom on London's fringes. In the City itself, these same enticements have additionally encouraged the City's emergence as the major control point for all sectors of the British economy. Today, around 60% of Britain's largest 500 industrial firms have their headquarters in the City (e.g. Evans, 1973; chapter 4). This role as a control centre is not restricted to London's national standing, for it has an international counterpart. Only New York stands above London with regard to its proportion of the headquarters of the world's largest 500 manufacturers. No other British city comes within the top 50 in this league (Feagin and Smith, 1987). Yet the critical point about these industrial concerns is that they are control points, not production points. Within London itself, manufacturing is a weak, declining economic sector (Chapter 3). Even including those in control functions, just over 19% of the City's workforce was manufacturing-based in 1981. The trend in this proportion is downwards, with business and financial services, which in 1981 accounted for 16.6% of workers, moving in the opposite direction (Buck *et al.*, 1986). The consequence is that a London-based, office-worker view of the country permeates corporate decision-making.

It is not simply in the private sector that a fixation on London is evident. In government it also makes itself felt with great clarity. In organizational terms, for one, London has repeatedly come in for 'special' treatment. Thus, the Metropolitan Police Force reports directly to the Home Secretary, whereas other forces are overseen by local institutions. Public transport has now joined it as a centrally manipulated institution (Chapter 8), while the peculiar character of the capital has also promoted a vast number of pseudo-governmental organizations covering functions commonly in local authority hands elsewhere. Cousins (1988), for example, suggests that there are well over 1000, and possibly over 2000, such institutions in the metropolitan area. In addition, London seemingly takes a disproportionate role in deciding the precise character of legislation. Thus, in the measures taken to evolve a national social policy in the early years of this century, it was the peculiar problems of the London labour market, most especially seen in the preoccupation of the Fabians with casual labour in the docks, that had the major formative role, even though the nation's key worker problems were in the North (Ashford, 1986). The need for legislative reform and for new policies have commonly been identified due to London's problems, and conceptualized for its special needs. Around London we find the first greenbelt. As a major governmental policy, new towns were introduced to address its particular problems. In 1965 Greater London became the scene of the first major local government reorganization this century (Chapter 12). Likewise, in seeking to 'regenerate' Britain's cities, the Thatcher government used the London Docklands Urban Development Corporation as the testing ground for its ideas and as the centrepiece of its propaganda campaign. If only because Members of Parliament and, more particularly the government (including its senior mandarins), live and work in London, the particular problems and trends of the capital impinge upon them most forcefully. Geographically, the capital is the common focus in their lives.

London as Diversity

The roles that London has fulfilled in national and international affairs have had a profound impact on its internal form and the lifestyles it supports. Here is a city of diversity. The contrasts which can be found are more extreme than those of most other cities, for overlaid on locally inspired social and geographical differentiations that are intrinsic to all cities are the intensifying and complicating forces of the capital's national and international roles. London is no New York, Rome or Tokyo, for these centres lack the primacy within their nations that London possesses; they have competitors for the attentions of their nation's elites. It is not even equivalent to Paris, for while these two cities share a common national standing (perhaps Paris is even more primate nationally), grafted onto London's position is a more encompassing international role (most especially in terms of finance and corporate control). Hence, the potential for internal differentiation is exacerbated.

Nevertheless there are clear dimensions to the areal divisions that exist. These do not fit into a simple theoretical model of urban structure. Trends manifest in present distributions of demographic, economic, social and political phenomena, as well as in the physical fabric of the city which their interactions produce, are commonly contradictory, invariably of dissimilar intensity, and generally cover different time spans (e.g. Chapter 9). Certainly, some broad spatial bands can be delimited which share common features. As early as the seventeenth century, for example, it was possible to distinguish the high status residential areas of the West End of London from the mercantilist City of London (Jones, 1980). Added to this still evident east–west divide has been a cleavage between the inner city and the suburbs. Of course, the suburbs of today are not those of yester-year. The inner ring of middle and higher income nineteenth century suburbs, comprising areas like Chiswick, Dulwich, Hampstead, and Herne Hill, are now more widely spaced on the social status ladder than when they were first built up (e.g. Chapter 10). Some have maintained positions of high standing, others are in the process of gentrification, while the physical fabric of some have a run-down, dispirited air. Yet even given the differences between them, these places can be clearly distinguished from the newer twentieth century suburbs of Beckenham, Enfield, Sutton and Wembley (as seen in other aspects of the city's socio-economic life; e.g. Chapter 3, Chapter 6 and Chapter 7). This last group is not homogeneous, for included in its midst were some large estates of low income public housing (Becontree, the largest and most renowned of which includes 25 000 homes and covers 3000 acres; see Garside, 1983). Yet within these newer suburban areas variety is less evident than in other districts of the capital. When it is present, it commonly arises from the existence of commercial or industrial activities. In some areas these cut across geographical dimensions of basic social divisions, in others they reinforce them. Swaths of industrial activity can be identified (Chapter 4), just as there is a hierarchy of central place centres for commerce and retailing within the metropolitan area. These provide no neat spatial patterning which urban modellers can drool over. Commercial centres like Bromley and Croydon are dotted across the southern metropolitan zone, but comparable places are less evident in the north (Chapter 7), and the industrial presence in western London has a quite different character from that along the Lea Valley or the eastern River Thames. There is a complexity here, one that cannot be understood simply in terms of national or international forces acting on London. For there is also a local dynamic to processes of change that both inspires differentiation between the capital and elsewhere, and induces differentiation within the capital itself (Chapter 11). It is the story behind this complexity which will be teased out in the chapters that follow.

Of course, spatial differentiation is not unchanging. The geography of London must be understood with reference to a matrix of space-time associations. What is more, at least since the passage of the London Green Belt Act in 1938, the physical fabric of this matrix has effectively had fixed external boundaries. Undoubtedly, this has placed restraints on the changes that have followed, but probably less than might be imagined. Infrastructures have been adapted to new demands made on them and behaviour patterns have been moulded to fit the physical limits they have faced. In addition, there has been no change in the importance of London for showing a lead for trends in British society as a whole. Due to the centrality of the capital in British life, when 'new fashions' start in London they have a greater chance of being identified and reported to the public at large. This is not simply because the media is centred there, but also because those who provide media channels with the bulk of their material – the government and the large corporations, for instance – also wear glasses with a heavy London tint. In politics, for example, the emergence of local authorities controlled by the new left has taken on a larger than life appearance in media presentations; a situation that cannot be divorced from the key role of London authorities within this movement (most especially given their conflicts with central government; Gyford, 1985). In addition, frequently as a by-product of the city's size, new dimensions to urban living are more easily identified and more easily encouraged here; and, as Feldman and Jones (1989) pointed out, this has a long history. In the case of neighbourhood gentrification, for example, London was a leader in enticing professionals back into the city centre (Glass, 1964). Certainly, the media played a part in establishing the social 'legitimacy' of gentrification trends. However, what should not be discounted as an aid to this redevelopment process was the availability of large tracts of housing with the potential for conversion. For buyers, this held out the prospect of not only gaining an improved house, but also of taking part in neighbourhood 'revitalization' (as professionals purchasing these homes and media commentators writing about them describe it at least). At the same time, for developers, there were advantages in that there was a possibility of attaining economies of scale in conversion practices (Smith and Williams, 1986). It is not that these trends were absent elsewhere in the country, but the cosmopolitan character of London's population, its significant wealth (and great diversity of wealth), and the sheer scale of problems that it has to face, mean that original trends are more likely to emerge here. More importantly, they are more likely to be identified, broadcast and accepted. What happens in London does not have to end up happening in Hartlepool, but what starts in Hartlepool has only a slight chance of becoming fashionable in the capital.

London as Research Site

Both the centrality of the capital to the British experience and its internal diversity and dynamism have had carry-over effects into research. London has proved a fertile ground for researchers. London life throws up innumerable questions, teases with imponderables, demands clarity where there is ambiguity and, through the sheer magnitude of events, thrusts research questions before the investigator with hints, prompts and screams for attention. It comes as no surprise to find that some of the classic studies of social science and history have been grounded in attempts to elucidate the character, performance, differentiations and challenges of this city. Whether it has been in searching out the processes of urban growth (Dyos, 1961), investigating poverty and homelessness (Greve *et al.*, 1971; Jones, 1971), tracing the intricate web of inter-family relations within neighbourhoods (Young and Willmott, 1964) or elucidating dimensions of local politics (Dearlove, 1973; Saunders, 1979), London-based investigations have stood amongst the forerunners in research on British society. Yet within the array of major contributions that have been made, it is notable that contributions from geographers have been less evident. Even within geography itself, London-based research has not charted a strong path which has provoked new theoretical perspectives, forged new conceptualizations or revealed previously unknown dimensions of society. A fundamental reason for this has been the immense diversity of the capital. As the chapters in previously published geographical reviews of London demonstrate, the geographical imagination stretches widely, embracing aspects of the life sciences, the physical sciences, the social sciences and the arts (see Clayton, 1964; Coppock and Prince, 1964; Clout and Wood, 1986). In part, the geographical perspective has sought an integration across these dimensions, while elsewhere it has been more concerned with understanding the geographical component of the events and processes others have concentrated upon more directly. Whichever route was chosen, the research process has been greatly complicated by the introduction into the analysis of an extra dimension (viz. space). In the main, the conceptual and theoretical armoury of the discipline has been too poorly developed to provide a sufficient arsenal to address this extra dimension adequately (this circumstance being shared with other disciplines, when they have addressed questions of geographical variability). However, it has been important to continue making the case for the need for a geographical dimension in societal investigations. Increasingly, this is being appreciated by those from other disciplines and appropriate concepts and ideas are being worked on (Giddens, 1984; Lash and Urry, 1987). There is still a long way to go. However, the time is ripe for, demonstrating once again, the need for understanding the spatial-temporal dimensions of London life. It is also evident that the contributions of the Centre for Urban Studies (1964) and Donnison and Eversley (1973) merit updating. It is in this spirit that this book has been written.

2

The Metropolitan Economy: Continuity and Change 1800–1939

David R. Green

Introduction: London and the Industrial Revolution

The industrial revolution, according to Hammond, was '... like a storm that passed over London and broke elsewhere' (Lee, 1986, 125). It was in the provinces, particularly Lancashire, the West Riding and the Midlands, that the new economic order was apparently being forged. Traditionally, historians have emphasized the role of the staple industries such as textiles, iron and engineering, together with mining, in explaining Britain's rise to industrial prominence (Cannadine, 1984). The factory rather than the workshop, and the mine rather than the counting house, were recognized as the foci for industrial development. Increases in output were allotted to steam and water-powered machinery rather than handicraft production. And it was in cities such as Manchester, Leeds, Birmingham and Liverpool, rather than London, that the industrial revolution gathered strength and broke with a ferocity that staggered both contemporaries and historians alike.

Recently, this conventional view has been questioned from two directions, both of which have implications for understanding the nature of London's economic development and its role in the national pattern of economic growth. First, the pre-eminence of factory production has been questioned. As late as 1851 more manufacturing employment occurred in workshops using simple hand tool technology than in more highly mechanized factories (Samuel, 1977). Even in textiles, handloom weaving continued to be important until well into the nineteenth century. In many trades, such as clothing and shoemaking, handicraft work competed successfully with factory goods throughout the 1800s (Bythell, 1978). That it was able to do so bore witness to the versatility and adaptability of small scale production, and in particular to the way in which improvements in productivity could be made without recourse to mechanization. Increasing the extent of subdivision and specialization, which in turn allowed the replacement of skilled by less skilled labour and the spread of piece work, were potent means of improving productivity within the confines of handicraft production.

Secondly, it has been argued that the importance of the service sector, particularly transport, finance and commerce, has been underestimated in relation to wealth creation. Between 1841 and 1911 jobs created in services accounted for over half the national increase in employment (Lee, 1986, 15 and 134). Of this, London and the South East accounted for at least half of new service jobs; that is 20% of national employment growth during the period (Lee,1986,

134–35). Taken as a whole, the service sector contributed more to Victorian economic growth than manufacturing, a situation confirmed by a far larger proportion of millionaires owing their fortunes to commerce rather than to industry (Rubinstein, 1977, 104; Lee, 1986, 11 and 15).

Both these qualifications have important implications for understanding industrialization in London and for placing the capital within a wider national framework of economic development. What is so striking about the capital's economy since 1800 has been its dynamism and versatility. Up until today, it has remained relatively unaffected by cyclical economic fluctuations and has consistently maintained a high rate of growth (Lee, 1986). To a large extent this has been due to the broad range of its manufacturing industries and to the importance of service employment. The service sector has traditionally been far less susceptible than manufacturing to cyclical fluctuations; thus providing a relatively stable employment base for the capital. This stability, coupled with the undoubted wealth of the London market, ensured that local demand for products remained high. In turn these factors underpinned, in the nineteenth century, the growth of traditional finishing trades, such as clothing, shoemaking and furniture, and in the twentieth century, the development of new consumer industries, such as electrical engineering and vehicle manufacture.

The history of London's economy is more than just of passing interest, for it touches upon the very nature of urban development and economic growth of the country as a whole. It emphasizes the role of the service sector in underpinning economic growth and questions the trajectory and form that industrialization took during the period. More specifically it draws attention to the tensions that exist between the city as a focus of consumption and as a centre of production. How these tensions were resolved provides the subject matter of this chapter. First, the role of demand is examined, emphasizing the great size and wealth of the London market. The focus then shifts from consumption to production and discusses the structural characteristics of the metropolitan economy. Although the economy was characterized by stability and continuity, nevertheless significant changes and realignments occurred. The processes underlying these changes in the nineteenth and early twentieth century are examined, emphasizing the relationships with and implications for the social geography of the city.

The London Market

Unlike the pattern of growth in regions dependent on export trades, economic development in London from 1800 relied heavily on the state of local demand. Londoners tended either to derive their incomes from services and distribution, which primarily served a local market, or from employment in construction and the finishing trades, which were similarly oriented towards local demand. The significance of London as a market for goods and services, therefore, is the starting point for understanding the pattern of economic development.

The size and wealth of the London market long predated the heyday of Victorian industrialization. Indeed, in seventeenth century Europe only Paris and Constantinople were larger (Beier and Finlay, 1986, 3–4). London's pre-eminence as the nation's capital, and indeed as a city of international repute, provided it with a unique position within the British economy. In 1650 its population was 375 000 but by 1801 it was approaching one million. The long upward trajectory of growth stemmed from the facts of national economic and political life. Ever since the fifteenth century, London had been the permanent home of mon-

Figure 2.1   Greater London in 1931

archy, court and parliament. Over time, the growth of state bureaucracy added layer upon layer of administrative, legislative and political activity in the capital, drawing those in search of political influence to the city. With the centralization of political life also came the centralization of economic activity. Not only did the aristocracy and gentry provide London with an ever increasing demand for luxuries and high quality goods but the concentration of governmental power facilitated access to finance and trade. The legal profession was similarly concentrated in the city and regulated the increasing volume of commercial transactions that passed through the capital. In turn the creation of wealth was reflected in the social and cultural life of the nation that centred on London (Fisher, 1948). It was here that music was composed, language created and plays performed. Artistic and scientific life revolved around the capital's academies and institutions. Social status and breeding were paraded in London's aristocratic clubs, whilst, in the mansions of the rich, partners for marriage were sought and won. Politics, economy, high society and culture were intertwined in mutually reinforcing relationships that ensured the continuation of London's pre-eminence.

The vigour and extent of London's growth during the nineteenth century, however, even took contemporaries by surprise. In 1852 George Sala wrote

'London has not grown in any natural, reasonable, understandable way ... it has swollen with frightful, alarming, supernatural rapidity. It has taken you unawares; it has dropped upon you without warning; it has started up without notice. . . .' (Dyos, 1982, 190). Not until 1851 did London become a full census division and not until then were its boundaries defined. By that time its population was 2 363 641 and in some directions the built-up area stretched more than five miles from the centre. But even this was dwarfed by subsequent growth. The boundaries defined in 1851 were by and large those adopted in 1889 by the London County Council (Figure 2.1). Even at that time London was spilling over its borders. By the turn of the century the idea of a 'Greater London' extending at least 15 miles from the centre became common. By 1900 it is therefore possible to speak of three distinctive geographical zones which comprised London: the old, inner core, the outer Victorian suburbs, both of which were included in the administrative county of London, and outer ring beyond the county boundary.

Table 2.1 Population of London 1801–1931

Year	Central area	Rest of London	Administrative County of London	Outer London	Greater London
1801	783000	176000	959310	155334	1114644
1811	892000	247000	1139355	184544	1323899
1821	1054000	325000	1379543	216808	1596351
1831	1219000	436000	1655582	247990	1903572
1841	1383000	566000	1949277	286067	2235344
1851	1560000	803000	2363341	317594	2680935
1861	1662000	1146000	2808494	414226	3222720
1871	1667000	1594000	3261396	624245	3885641
1881	1648594	2181703	3830297	936364	4766661
1891	1579261	2648693	4227954	1405852	5633806
1901	1529136	3007131	4536267	2045135	6581402
1911	1393013	3128672	4521685	2796730	7251358
1921	1266843	3217680	4484523	2995678	7480201
1931	1169000	3227000	4396821	3805997	8202818

Source: London County Council, London Statistics, volume 35 (1930–31, 23)
Note: see Figure 2.1 for geographical definition of areas

Massive population growth up to 1851 filled the old core and overflowed into surrounding districts (Table 2.1). From mid-century, however, the City of London and surrounding inner districts began to lose population and housing as a result of demolitions for warehouse construction, street clearances and railway building (Jones, 1971). Those who could afford to use omnibuses and railways left for the suburbs, whilst the remainder were squeezed into already densely settled inner neighbourhoods. In such districts overcrowding increased substantially, resulting in tremendous pressures on housing and encouraging the rapid formation of slums (Wohl, 1977; Green and Parton, 1990). Whilst large landed estates in western areas were mainly able to keep the tide of poor from breaking through, the absence of such institutional landlords in eastern and, to some extent, southern districts meant barriers were more difficult to erect. It was into these districts that the displaced poor were squeezed whilst the wealthy beat a path to the West End and suburbs.

This situation continued until the 1870s but from then pressures on housing in central areas caused an absolute decline in population. Much of the outflow was absorbed by suburban growth within the administrative county of London. From the 1880s the outer districts contained a larger number of people than the older, inner core (Table 2.1). This shift presaged an even more striking change in demography. From the 1880s the pace of suburbanization accelerated sharply as cheaper and more efficient means of transport were introduced. In 1849 less than 30 miles of railway existed in London but by 1880 this had risen to over 215 miles and by 1900 over 250 miles of track were in operation. The wide-spread introduction of workmen's fares from the 1880s allowed the expansion of working-class suburbanization. The underground railway, first started in 1854, provided an additional option for movement within the inner ring and by the 1870s had spread as far as Richmond. Shorter journeys were made possible by the development of horse drawn omnibuses and trams which served localities closer to the centre. Competition between tramway companies, coupled with the London County Council's policy, late in the century, of reducing fares on public transport, helped lower the cost of commuting and further encouraged suburbanization. The rise in the number of journeys, from 256.4 million in 1880 to 819.2 million in 1900, bears witness to the impact of transport improvements.

The result of these changes was a fundamental redrawing of the demographic geography of London. As shown in Figure 2.1, whilst most districts in the inner ring lost population, spectacular growth occurred in several outer districts, notably Southgate to the north, Croydon and Merton in the south, Southall and Wembley to the west, and Chingford and Ilford in the east. The outer ring as a whole gained nearly 685 000 persons between 1901 and 1911 whilst for the first time the County of London lost population. The First World War called a temporary halt to the suburban rush but in the decade that followed decen-tralization again gathered pace. Aided by the spread of car ownership and the introduction of motorized bus services, and coupled with improvements in both the carrying capacity and speed of public transport, the outer ring gained 810 319 people in the years 1921–31. By the latter date the population of the inner and outer rings was evenly balanced, and together they comprised a population of over eight million people, nearly one in five of the total British population.

But this was only part of the story. Urbanization itself stimulated demand by creating a market for a variety of products, services and amenities, such as processed food, transport and distribution and cleaning services, not to mention entertainment and leisure, that would otherwise not necessarily have been required. Furthermore, suburbanization was important in generating higher levels of demand for an entirely new infrastructure both inside and outside the home (Walker, 1978; Harvey, 1985, 122). Spatial expansion of the suburbs pro-vided demand for social amenities, such as roads, shopping centres, hospitals and schools and added to those employed in transport, retail and distribution. Within the home, the decline of domestic servants from the start of the twen-tieth century, coupled with the provision of electricity, provided the basis for a sharp rise in the demand for household appliances like vacuum cleaners, food blenders, washing machines and fridges. Indeed, by creating an effective demand for such products, suburbanization during the inter-war period helped to sustain high rates of growth in new industries in contrast to the experience of depression elsewhere.

Size and suburbanization, however, were not the only factors that underpinned the continued dynamism of London's economy. In addition to being the largest concentration of people, London's wealth set it apart from other areas of the country. As the hub of the empire and seat of government, the capital maintained its hold on the Victorian and Edwardian elite. Not only were the very wealthy

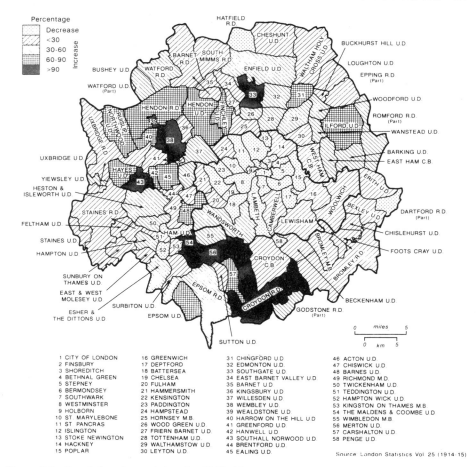

Percentage

Decrease
<30
30-60
60-90
>90
Increase

Figure 2.2   Population change in London 1901–11

1 CITY OF LONDON	16 GREENWICH	31 CHINGFORD U.D	46 ACTON U.D.
2 FINSBURY	17 DEPTFORD	32 EDMONTON U.D	47 CHISWICK U.D.
3 SHOREDITCH	18 BATTERSEA	33 SOUTHGATE U.D	48 BARNES U.D.
4 BETHNAL GREEN	19 CHELSEA	34 EAST BARNET VALLEY U.D	49 RICHMOND M.D.
5 STEPNEY	20 FULHAM	35 BARNET U.D	50 TWICKENHAM U.D.
6 BERMONDSEY	21 HAMMERSMITH	36 KINGSBURY U.D	51 TEDDINGTON U.D.
7 SOUTHWARK	22 KENSINGTON	37 WILLESDEN U.D.	52 HAMPTON WICK U.D.
8 WESTMINSTER	23 PADDINGTON	38 WEMBLEY U.D.	53 KINGSTON ON THAMES M.B.
9 HOLBORN	24 HAMPSTEAD	39 WEALDSTONE U.D	54 THE MALDENS & COOMBE U.D
10 ST. MARYLEBONE	25 HORNSEY M.B.	40 HARROW ON THE HILL U.D	55 WIMBLEDON M.B.
11 ST. PANCRAS	26 WOOD GREEN U.D	41 GREENFORD U.D	56 MERTON U.D.
12 ISLINGTON	27 FRIERN BARNET U.D.	42 HANWELL U.D	57 CARSHALTON U.D.
13 STOKE NEWINGTON	28 TOTTENHAM U.D.	43 SOUTHALL NORWOOD U.D.	58 PENGE U.D.
14 HACKNEY	29 WALTHAMSTOW U.D	44 BRENTFORD U.D.	
15 POPLAR	30 LEYTON U.D.	45 EALING U.D.	

Source: London Statistics Vol 25 (1914-15)

concentrated in London but its middle class was richer and more numerous than in the provinces (Rubinstein, 1977, 111). Together, these groups provided the demand, not only for domestic servants and high quality housing, but also for a large and stable market for bespoke trades, such as tailoring, shoemaking and furniture in the West End (Hall, 1962).

Of more specific relevance to London's manufacturing were changes from the 1870s that witnessed the creation of a mass market for consumer goods. This came about largely due to changes in working class consumption, prompted by rising real wages and a decline in family size. Between 1870 and 1935, with the exception of the immediate pre-war period, the level of real wages in London more than doubled, prompted primarily by falls in the price of food (Tucker, 1936). Other factors contributed to the rise of a mass market for consumer goods. First, significant reductions occurred in the size of families, with the sharpest falls occurring first amongst professional groups and white collar workers but spreading from there to include working class households (Baines, 1981). In mid-Victorian England the average number of children per family lay between 5.5 and 6.1 but by the 1920s this had fallen to 2.2 (Thane, 1981, 356). Reductions of a similar magnitude were evident in London where by 1931 average family size had fallen to 3.46 persons. Given that Charles Booth in the 1890s had blamed large families for approximately 10% of those in poverty in London, reductions in family size had the effect of increasing considerably the number of households with excess disposable income (Booth, 1902a, 147).

Secondly, over time the proportion of the population in paid work rose which increased the number of wage earners per household, thereby expanding the potential size of the market. In London in 1891 the activity rate for males and females was 62% and 29% of the population respectively, but by 1921 these percentages had risen to 67% and 31% (Llewellyn Smith, 1934a, 327). The entry of more women into paid employment often added a second household income. Moreover, there was a tendency for workers to enter more secure and better paid jobs, notably commercial, clerical and other white collar employment. During the nineteenth century wages for these jobs rose three times as fast as for manual work (Lee, 1986, 139). Women in particular were moving into new areas of employment, away from domestic service and into offices, shops and assembly work in the new manufacturing industries (Alexander, 1989). The growth of household incomes, therefore, both in terms of larger numbers in paid employment as well as in better jobs, was of prime importance in stimulating consumer demand.

The outcome of these interrelated factors was clear. As a centre of government and finance a whole army of brokers, factors, clerks, civil servants, lawyers and soldiers was spawned; as a port for the empire and transport centre for the country, not to mention a city of enormous areal extent, thousands were employed in the movement of people, goods and messages; and as the largest and wealthiest market in the country, consumer demand stimulated growth in manufacturing geared towards the production of finished items for local consumption. Furthermore, the size and stability of the metropolitan economy ensured that demand for consumer products remained relatively buoyant. Proximity to the market was crucial for a host of traditional trades, such as printing, clothing, shoemaking and furniture, as well as for several new industries, such as food processing, electrical engineering and vehicle manufacture, in the twentieth century. An appreciation of the significance of local demand in stimulating metropolitan industries is therefore essential for a full understanding of the dynamism and vitality of the London economy in the past two centuries.

The London Economy

What is perhaps so striking about the London economy is not so much its novelty as its consonance with patterns in the past. From as early as the sixteenth century London's growth depended on trade and commerce. Descriptions of the city at the time included accounts of its mercantile and commercial prowess. In 1592 Frederick, Duke of Wirtemburg, described it as '... a large, excellent, and mighty city of business, and the most important in the whole kingdom; most of the inhabitants are employed in buying and selling merchandise, and trading in almost every corner of the world...' (Manley, 1986, 35). During the eighteenth century British expansion led to the growth of overseas trade, much of which was organized and financed by London merchants and routed through the port. In turn the wealth accumulated through such trade was often expended in the capital, thereby laying the basis for the development of a variety of high quality manufacturing trades. As Daniel Defoe wrote in 1727, 'London consumes all, circulates all, exports all, and at last pays for all, and this greatness and wealth of the City is the Soul of the Commerce to all the nation' (Fox, 1987, 94).

In the course of the nineteenth century these trends continued. Unlike other industrial cities of the time, London's growth was never led by manufacturing but by services and transport. Although it contained a vast and varied manufacturing base and was the greatest centre of industrial production in the country, at no time during the nineteenth century did industry provide the bulk of its employment (Table 2.2). Indeed, services and transport accounted for over half the workforce throughout the period examined here, with manufacturing accounting for most of the remainder. Over time very little change occurred in the sectoral distribution of the workforce and at this level of generalization, therefore, the economy was characterized by a high degree of stability.

Table 2.2 Employment in Greater London 1861–1921

Sector	1861		1921	
	Number (000)	Per cent	Number (000)	Per cent
Services	666.7	45.1	1596.0	49.6
Manufacturing	468.8	31.7	1052.6	32.7
Transport	138.2	9.4	346.6	10.8
Construction	98.1	6.6	146.9	4.6
Others	107.0	7.2	74.0	2.3
Total	1478.8	100.0	3216.1	100.0

Source: Hall (1962)

Note: Services include gas, water and electricity, distribution, insurance, banking and finance, public administration and defence, professional occupations and miscellaneous services.

The range of manufacturing trades in London was enormous but typically industries were oriented towards the finishing trades, characterized by a high value in relation to bulk and dependent on close contact with the final consumer. During the nineteenth century, clothing, shoemaking, furniture, printing, metals and engineering, and precision manufacture were amongst the most important trades (Table 2.3). In the twentieth century other consumer industries, such as electrical engineering and vehicle manufacture, grew in importance whilst more traditional trades declined. In both cases, proximity to the London market was of overriding importance. In contrast, London was deficient in trades that were heavily reliant on raw materials and in the production of semi-finished goods. Industries such as textiles, leather, heavy engineering and shipbuilding left the capital altogether whilst others, such as chemicals and rubber, moved to sites in more peripheral locations (Martin, 1966; Jones, 1971, 19–26).

A further characteristic of London's manufacturing was the absence of a large scale factory system. Small workshop production was dominant. Distance from supplies of raw materials, the high cost of fuel and shortage of space all inhibited the development of the factory system. Some large employers did exist, notably in brewing, construction, engineering, shipbuilding and transport, but by and large these were exceptions rather than the rule. Only in southern riverside districts and in peripheral localities, such as around the docks in Poplar and West Ham, were factories of any importance. Elsewhere small workshops and domestic industry dominated. As the Registrar General noted in 1851, '. . . the most impressive feature of industry is not that the few are so large, but that the many are so small'. (British Parliamentary Papers, 1852–53). According to 1851 census figures, 86%

Table 2.3   Occupational Structure of the County of London 1851–1931

Males	1851		1901		1931	
	Number	Per cent	Number	Per cent	Number	Per cent
Construction	61319	10.2	149962	10.7	101384	7.9
Conveyance	59998	10.0	243924	17.4	158414	12.3
Food and drink manufacture	59844	10.0	138762	9.9	46240	3.6
Clothing, boot and shoe making	59235	9.9	81187	5.8	58874	4.6
General labour	57248	9.6	75010	5.4	NA	NA
Professional	38519	6.4	65407	4.7	47643	3.7
Metal and engineering	31392	5.2	95503	6.8	123270	9.6
Wood and furniture	29214	4.9	61891	4.4	42704	3.3
Domestic service	25660	4.3	53525	3.8	105123	8.2*
Commerce, finance	22198	3.7	134261	9.6	298218	23.1
Local and central government	18960	3.2	46638	3.3	124520	9.7
Printing and paper	18746	3.1	63566	4.5	62245	4.7
Others	116442	19.5	190333	13.6	123224	9.6
Total	598775	100.0	1399969	100.0	1289859	100.0

Females	1851		1901		1931	
	Number	Per cent	Number	Per cent	Number	Per cent
Domestic service	171123	52.0	328337	45.6	250792	31.6*
Clothing, boot and shoe making	89908	27.3	156050	21.7	113996	14.4
Professional	18212	5.5	52962	7.4	10194	1.3
Food and drink manufacture	14819	4.5	49492	6.9	39568	5.0
Commerce, finance	3233	1.0	20285	2.8	143606	18.1
Printing and paper	3164	1.0	33369	4.6	36703	4.6
Local and central government	724	0.2	5796	0.8	51196	6.5
Metals and engineering	494	0.2	3932	0.6	29564	3.7
Others	27194	8.3	69108	9.6	118571	14.9
Total	328871	100.0	719331	100.0	794190	100.1

Source: Census for England and Wales, 1851, 1901, 1931.

Note: *Includes employment in hotels and lodging houses.

of employers had less than ten workers and nearly 70% employed less than five men. Small scale production had several clear advantages in London. First, many trades required little capital to start up therefore barriers to entry were low. Secondly, small entrepreneurs maintained close contact with the market and were able to respond quickly to changes in demand. As Charles Booth noted: '. . . the experience of London affords practical proof of the persistent vitality of small methods of business. . .' (Booth, 1903, 70).

In another sense London's economy was also characterized by stability. Since at least the 1850s growth in London and the South East was faster and less subject to fluctuation than elsewhere in the country, a situation which has continued until the present day (Lee, 1986; Marshall, 1987). As a result of the importance of services and its diverse manufacturing base, unemployment in London tended to remain below that for the rest of the country. This applied even in the nineteenth century, although many trades were notoriously seasonal and casual work, especially in the docks, was a peculiarly intractable problem (Southall, 1988). During the inter-war period the distinctiveness was even clearer. Buttressed by the size of the service sector and by virtue of a higher proportion of new and expanding industries, unemployment rates in London and the South East were consistently lower than those of other industrial regions. In 1932, for example, at the height of depression, the employment rate in the country as a whole was 22.1% whilst in London it was 13.5%. Even in more prosperous years, London rates compared favourably with the rest of the country. So in 1937, when the national unemployment rate was 10.8%, in London it was only 6.3% (Beck, 1951; Crafts, 1987; Ward, 1988).

However, despite this appearance of stability major changes and realignments did occur in London's economy. Whilst the broad framework of employment did not fundamentally alter there were significant changes in specific sectors as new occupations emerged and traditional ones declined (Table 2.3). Thus, over the period as a whole, clothing and shoemaking declined in importance, being beaten down by cheaper provincial competition, and some smaller trades, such as tanning and typefounding, found London costs prohibitive and sought out provincial locations in which to continue production. On the other hand, employment in metal and engineering increased, particularly from the 1900s when mechanical and electrical engineering emerged as leading sectors in the economy. Major changes occurred in female employment. The decline in domestic services had major consequences for women, as did the demise of clothing, although many more found employment in clerical jobs, as teachers and as shop assistants. The First World War brought other changes in female employment, as women entered trades traditionally monopolized by men, such as armaments, engineering, ship repair, gas stoking and brewing (Thom, 1989). After the War, much light assembly work in the expanding field of electrical engineering was taken up by women. The expansion of commerce, retailing and local government also provided many new opportunities for employment.

Other changes were also important. In manufacturing, the stimulus that wartime production gave to standardization and mass production, and the impact of electricity on freeing industries from reliance on coal, resulted in the spread of mechanization, an increase in semi-skilled and unskilled work and the development of new factories in the outer districts. Despite these changes, however, industry remained geared to providing goods for the London market. That it did so in different buildings located in new areas and with a different workforce was important but it should not blind us to the fact that the London market still exerted the most powerful force on the development of manufacturing in the capital.

Nineteenth-century London: Social Geography and Industrial Change

In the first half of the nineteenth century London's economic geography crystallized around three complementary but distinctive zones. At the centre the City of London was a clearly defined commercial and financial core containing a mix of offices, commercial premises and warehouses. Although the full impact of depopulation had yet to be felt, by 1850 wealthier residents had already begun the suburban trek. The outer suburban zone consisted of the

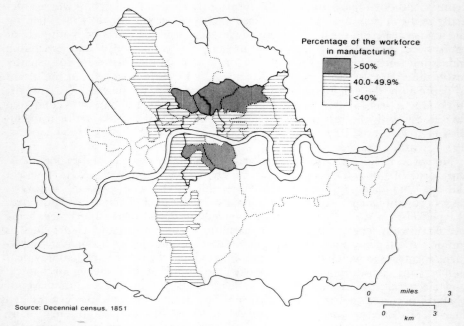

Source: Decennial census, 1851

Figure 2.3   The inner industrial perimeter 1851

homes of the professional and commercial middle classes, together with their retinues of domestic servants. Sandwiched between core and suburb was the inner industrial perimeter stretching in a belt around the City of London, from Holborn in the west to Poplar in the east and southwards to Lambeth. This contained the bulk of manufacturing employment (Figure 2.3). Many of London's traditional trades were concentrated in this inner perimeter, and on their success depended the livelihoods of large numbers of workers. Other trades, such as printing, watchmaking, silk weaving and engineering were also concentrated in particular localities within the inner zone (Spate, 1938).

In the course of the nineteenth century pressures mounted on the viability of many manufacturing trades in the capital, resulting either in a restructuring of production or in a relocation of activity. The cost of space in London, both in terms of land values and rates, was a pressing problem. During the 1870s an acre of land in the City of London was valued at about £14 520 whilst a suburban acre was about £726, a ratio of about 20:1 (Dyos, 1961, 209: Wohl, 1977, 286). By the early 1900s competition from commerce and finance had pushed prices up considerably beyond this level. One plot near the Bank of England, for example, was sold in 1905 for £3.25 million an acre, which, although exceptional, was indicative of the intense pressure on land prices and site costs in the core (Offer, 1981, 255). Nor was this pressure confined to the City of London alone, for high land values spilled out into surrounding districts. During the 1930s, the cost of land in Finsbury, for example, was approximately £38 000 per acre compared to about £1000 in the outer suburbs (Yelling, 1990).

In addition to rents, from the 1850s property rates became a growing burden for businesses in London (particularly those in poorer localities) as property taxes fell heavily on industry and commerce. In 1897 it was estimated that whilst 9% of householder income was taken up by such taxes, this figure rose to 12% for shopkeepers and 14% for builders and factory owners (Davis, 1988, 37). Rising rate levels were the result of two processes. First, the outmigration of relatively wealthy middle class ratepayers from central, eastern and inner-southern districts from mid-century, coupled with the influx of a poorer population arising from demolitions in the centre, eroded the fiscal basis of many districts at the same time as the demand for poor relief was rising. This imbalance in local finance worsened sharply from the 1840s and, ultimately, was the reason for the establishment of the Metropolitan Common Poor Fund in 1867. Secondly, the creation in 1855 of the Metropolitan Board of Works, followed by its successor, the London County Council (LCC) in 1889, led to increased costs in order to provide infrastructural improvements. The LCC, in particular, became involved in a wide range of improvements, including an ambitious housebuilding programme, that resulted in large increases in property taxes (Offer, 1981, 166-67; Davis, 1988, 144–47 and 154–55). The LCC was also responsible for the enforcement of increasingly stringent controls on noxious trades and manufacturing workshops, which encouraged firms to relocate to areas of laxer control beyond its juris-diction (Howarth and Wilson, 1907, 145–47; Martin, 1966, 10).

The pressures to relocate production were felt by several trades, primarily by those which made large demands on space. As Charles Booth noted:

> The grouping on the outer ring of London of manufactures, such as of soap or chemicals, that require relatively to the numbers employed, large premises, may be necessary in conformity to municipal regulations but is likewise explained by the double necessity of avoiding high rentals and of securing the early command of cheap means of transport (Booth, 1903, 99).

Large users of land, such as the dock companies and railways were most seriously affected by spiralling rates and high site costs but were tied inextricably to London. In contrast, from the 1890s onwards trades which could relocate frequently did so, either to peripheral districts, such as West Ham or the Lea Valley, or to locations further afield. Such moves occurred with increasing frequency from the turn of the century. Thus, Napiers moved their engineering works from Lambeth in 1904 to a new site in Acton with room for expansion, whilst Yarrow Shipbuilders in Poplar cited high property taxes as the reason for their move in 1906 (Offer, 1981, 289). Siemens were prevented by LCC regulations from building an electrical machinery plant close to their cable works at Charlton and instead chose Stratford (Martin, 1966, 103). And even small workshops were not immune from stricter regulation with the enforcement during the 1890s of the Factory Acts in clothing and shoemaking encouraging the spread of outwork into districts with laxer controls (Schmeichen, 1984, 140–47).

Relocation, however, was less important in the finishing trades, such as clothing, shoemaking and furniture making, in which the major proportion of production costs fell to labour. With wage levels higher in London than elsewhere, competition from the provinces and abroad was acute, particularly once the railway system had brought major towns within easy reach of the city. From the 1850s, for example, wholesale clothiers in Leeds competed directly with those in London (Morris, 1986). For employers in the capital these pressures could only be met by reducing labour costs through 'sweating' the workforce. (No precise definition of sweating existed but it was generally a system whereby work was subdivided and subcontracted, with cheap, unskilled labour being employed for excessively long hours in insanitary homes and workshops; Morris, 1986, 8.) By breaking down work into uniform, component tasks it became easier to institute piece payment as opposed to time wages. This alteration obviated the need for supervision and allowed production to be removed from employers' premises to workers' homes, thereby allowing economies on the cost of light, heat and rent.

It was in London, more so than in any other place, that this system took firm hold. No finishing trade, indeed, not even construction, escaped the pressures of sweating. From the 1830s, as economic downturn encouraged manufacturers to reduce costs, and as the market for cheaper, ready-made clothing began to develop, wholesale clothiers, or shopsellers, began to extend the subdivision of work and expand the proportion of outwork. Shoemaking and furniture followed a similar pattern, and again it was local demand for ready-made products that forced the change. The introduction of the sewing machine and the band saw from mid-century greatly speeded up production in these trades, but mechanization was the outcome and not the cause of the subdivision of labour that prevailed. In clothing and shoemaking, the sewing machine made possible a massive increase in the rate of output whilst at the same time reducing labour costs; as such, it allowed London employers to meet the growing demand of the mass market in the 1870s. Cut-throat competition, both within London and from provincial manufacturers, further added to the pressures of sweating, a situation thrown into sharp focus during the 1880s and 1890s as depression undermined the metropolitan economy (Morris, 1986).

It was through the social geography of London that these pressures on employment were translated into patterns on the ground. According to Hobsbawm (1964) London comprised not one but three distinctive labour markets; the West End was dominated by bespoke production carried on by skilled, unionized artisans; the East End, in which manufacturing was for the wholesale trade, was dominated by unskilled and poorly organized workers; and the south which stood between the two. The fragmented nature of small-scale production, coupled with the short-term immobility of labour, ensured that these labour market districts were distinct entities. These spatial divisions within London were important, particularly in relation to trade unionism and the impact of labour militancy on production processes. The strength of unionism in the skilled trades acted as a significant brake on the employment of less skilled labour and the spread of cheaper forms of production. Tailors in the bespoke sector, for example, resisted the spread of outwork and subcontracting until 1834, at which point they could hold out no longer against the flood of unskilled production (Parssinen and Prothero, 1977). Where trade unions were weak, barriers to innovations in the labour process were correspondingly weaker. The sewing machine, for example, was first introduced in the 1850s to shoemaking and clothing in the unskilled, wholesale sector and spread from there into the more highly skilled and organized sector of production. Trade unionism throughout the eastern districts was generally weaker than elsewhere, primarily as a result of the dominance of unskilled, casual work and the large numbers of domestic workers isolated within their own homes. Not until the 'new unionism' of the 1880s and 1890s did trade unions spread beyond a narrow craft basis to incorporate significant numbers of unskilled workers (Hobsbawm, 1984). Significantly it was in eastern districts, such as Poplar and West Ham, which contained large factories, gas works and docks, that this unskilled, mass unionism made its greatest impact (Bush, 1984; Gillespie, 1989).

Clothing and shoemaking conformed closely to this broad pattern and illustrate the distinctions that existed. The West End and East End consisted of contrasting, and in some respects competing, areas of manufacture. 'Honourable' production for the bespoke market, undertaken by skilled, unionized male artisans employed in masters' premises was firmly concentrated in the western districts. In contrast, the 'dishonourable' trade involving cheaper, unskilled workers and frequently using the labour of wives and children employed in their own homes, was found throughout the eastern districts (Hall, 1962). The distinction between the two, according to Henry Mayhew, extended beyond the individual trades and marked a fundamental distinction in the outlook and habits of workers in the capital: 'In passing from the skilled operative of the West-end to the unskilled workmen of the eastern quarter of London the moral and intellectual change is so great, that it seems as if we were in a new land, and among another race' (Mayhew, 1981, 150).

In the course of the nineteenth century bespoke production of clothing and shoemaking in the West End continued but declined in importance. Instead, as the location quotients contained in Table 2.4 indicate, eastern districts specializing in cheaper quality products captured the main, ready-made market that developed as the period wore on. From at least the start of the nineteenth century manufacturing in the eastern districts concentrated on cheaper, less skilled forms of production. Clothes and shoes made in the district were primarily for the wholesale trade, the army, police force or the colonies. A relatively large and poor labour force, comprising a high proportion of women, was available. Without a resident middle class in the area the demand for domestic servants was low and this lack of alternative employment, coupled with the insecure earnings of menfolk employed in the docks, ensured that women were forced to enter the ranks of sweated labour. The mass influx of Jews to Stepney and Whitechapel from the 1880s added to the pool of labour but of itself did not create the conditions for sweating. As the century wore on cheap production expanded in eastern districts such as Shoreditch, Bethnal Green and Stepney, and came to dominate clothing, shoemaking and furniture manufacture.

Changes in the economic geography of London involved more than just the dispersal of some industries to peripheral locations and the expansion of sweated production in eastern districts. Other sectors of employment, such as the professions, clerical work and construction, also had distinctive geographies. In 1851, as Table 2.4 suggests, the capital could be divided into three broad zones. Western, northern and southern districts tended to have concentrations of building workers, of professionals and of white collar employees. With the exception of metal and engineering in Lambeth, Southwark and Greenwich, manufacturing was conspicuous by its absence. In contrast, eastern districts had a surfeit of industrial workers and those involved in transport at the docks, but were deficient in domestic service, the professions and white collar employment. Central districts, including the City of London itself, maintained a broad economic mix.

By 1891, however, the distinctiveness of each region had been accentuated, a fact reflected in the growing separation of the classes within Victorian London (Jones, 1971). The suburbs strengthened their hold on jobs in construction, the professions and white collar occupations. With the exception of metal work and furniture, central districts lost their manufacturing base. The demise of shoemaking and clothing, driven out by competition for space from commerce, involved the relocation of thousands of workers. To a certain extent both trades spread from there into the inner parts of the northern suburbs of Islington and St Pancras. More significantly, however, they became concentrated in the East End, which over the period became more solidly industrialized and more solidly working class, thereby laying the basis both for new unionism and working class radicalism in the 1890s and 1900s (Gillespie, 1989, 165).

Taken as a whole, these changes in the economic structure of London were important in terms of the geography of class separation but probably accentuated

Table 2.4 London 1851 and 1891: location quotients by occupational group

Occupation	West		North		Central		East		South	
	1851	1891	1851	1891	1851	1891	1851	1891	1851	1891
Professional	1.17	1.27	1.40	1.20	1.10	0.79	0.48	0.38	0.89	1.05
Commerce and finance	1.03	0.89	1.35	1.28	0.85	0.90	0.61	0.40	1.11	1.17
Clerical	0.73	0.82	1.27	1.31	1.13	0.72	0.67	0.52	1.14	1.17
Retail and distribution	0.62	0.87	0.83	1.01	1.20	1.07	1.30	1.01	1.04	1.05
Service	1.60	1.62	1.31	1.01	0.81	1.08	0.49	0.51	0.87	0.86
Building	1.14	1.11	1.25	1.03	0.72	0.62	0.79	0.72	1.06	1.14
Clothing	0.84	0.96	0.93	1.05	1.18	0.98	1.19	1.44	0.89	0.80
Shoemaking	0.74	0.55	0.87	0.95	1.17	0.66	1.33	2.54	0.90	0.61
Metal and engineering	0.68	0.57	0.70	0.79	1.19	1.06	1.13	1.15	1.24	1.31
Furniture	0.56	0.56	0.84	1.16	1.09	1.08	1.56	1.90	0.93	0.68
Transport	0.70	0.90	0.77	0.91	1.01	0.96	1.54	1.31	0.96	0.97

Source: 1851 census, Jones (1971)

Note: West – Kensington, Chelsea, St George's Hanover Square, Westminster, St James'

North – Marylebone, Hampstead, St Pancras, Islington, Hackney

Central – St Giles, Strand, Holborn, Clerkenwell, St Luke's, East London, City of London, West London, St Martin's

East – Shoreditch, Bethnal Green, Whitechapel, St George's-in-the-East, Stepney, Mile End, Poplar

South – St Saviour's, St Olave's, Bermondsey, St George's Southwark, Newington, Lambeth, Wandsworth, Camberwell, Rotherhithe, Greenwich, Lewisham

The location quotient is a measure of the concentration of a particular occupation in a particular district. It is calculated for each district by dividing the percentage of the workforce employed in a specific occupation by the percentage of the total working population in that district. For example, in 1851 West London contained 17.4 per cent of the working population in London. It contained 20.3 per cent of those employed in the professions. The location quotient is therefore:

$$\frac{20.3}{17.4} = 1.17$$

If the location quotient exceeds 1 it means that the occupation in question is over-represented in the district. A value below 1 indicates a degree of under-representation.

Table 2.5 The distribution of social classes according to Charles Booth, 1891

District	Class (per cent of population)			
	A and B	C and D	E and F	G and H
West	5.0	19.1	48.5	27.4
North	7.8	19.7	51.0	21.5
Central	15.0	24.9	48.0	12.1
East	13.2	25.0	56.8	4.9
South	7.1	23.7	51.3	17.9
London	8.4	22.3	51.5	17.8

Source: Booth (1891, 61–2)

Note: Booth's classification of social classes referred to the following description: A and B – lowest class and the very poor; C and D – poor; E and F – working class, comfortable; G and H – lower and upper middle classes and above. See Booth, first series, Poverty, volume 2, 20.
For a description of districts see Table 2.4.

trends already in existence (Green, 1985). Concern over class separation in Victorian London heightened with some justification as the period progressed (Jones, 1971). In eastern districts high property taxes and the spread of industry into residential neighbourhoods drove remaining middle class residents into adjoining suburbs. As Table 2.5 illustrates, what distinguished the eastern districts at the time of Charles Booth's poverty survey was less the presence of a large proportion of the poor than the absence of a substantial middle class element. Indeed, absolute poverty, if anything, was worse in the congested central districts where high rents squeezed the poor into isolated pockets of intensely overcrowded and insanitary housing (Valpy, 1889). Thus, whilst western districts became more solidly middle class and lost their artisan connections, the East End became a quintessentially working class region with a reputation for economic, political and cultural distinctiveness based around manual employment, mean streets and monotony (Keating, 1971, 103–24).

Twentieth-century London: Social Geography and Industrial Change

From the turn of the century, and particularly after the First World War, the British economy underwent a thorough transformation in terms of the nature of industrial activity and the location of economic growth. Following the First World War the decline of the traditional export trades of textiles, heavy industry and mining cast a shadow of depression across entire regions which during the Victorian period had experienced rapid growth. As a result, during the inter-war period, the North West, North, Wales and Scotland, in which such industries were concentrated, experienced high rates of unemployment and outmigration. In contrast rapid rates of growth occurred elsewhere, notably in London and the South East, where services and new consumer industries catering for domestic demand, were concentrated (Pollard, 1983, 76–82; Lee, 1986, 259–65; Ward, 1988).

The growth of London during the inter-war period has already been alluded to but it is worthwhile reiterating the broad dimensions of change. Between 1911 and 1939 the population of London increased from 7.25 to 8.73 million, with the entire increase taking place in the outer ring beyond the LCC boundaries (Table 2.1; Young and Garside, 1982, 342). By the latter date nearly one in five people in Great Britain lived in London and over half of these lived in its outer districts. This massive suburban growth was facilitated by major changes in the pattern of transport, relating first, to the expansion of the transportation system, secondly to the frequency of travel, and thirdly to the form of transport. Between 1902 and 1928 there was a 352% increase in the number of passengers carried by public transport in London (Llewellyn Smith, 1934a, 194). Furthermore, over the same period the number of journeys per person by public transport increased from 166 to 496 per annum, reflecting much higher levels of personal mobility. Finally, as Table 2.6 shows, there was a fundamental shift in the mode of travel from rail to road, with bus travel taking the lion's share of the increase (Dyos and Aldcroft, 1974, 374–79). In addition to the use of public transport, private car ownership within the country rose substantially from 32 000 vehicles in 1907 to over 1 800 000 in 1938, with London and the surrounding counties accounting for over a quarter of the total (Chisholm, 1938, 53–55; Burnett, 1986, 257). Large arterial roads built around the edge of London after the First World War, such as the Western Avenue, North Circular Road and the Great West Road, also facilitated greater levels of mobility and opened new areas for development. The Great West Road, for example, which opened in 1925, soon became known as 'the Golden Mile' by virtue of the large number of new factories built along its route, including several belonging to American companies such as Firestone Tyres, Gillette and Hoover (Weightman and Humphries, 1984, 58).

Table 2.6   Public transport in Greater London, 1900–29

	1900	1929
	(millions of passengers)	
Tramway	340.2	1076.3
Omnibus	264.5	1912.1
Railway and underground	214.5	648.8

Source: Llewellyn Smith (1934a, 174–5)

Improved roads and public transport, coupled with easier access to housing finance and the continued upward trend in real wages, helped to create a mass market for suburban homes. Supply was met from two sources, with private builders providing about three-quarters of new housing during the inter-war years, particularly from 1928 onwards. The LCC and local authorities provided the remainder, although throughout the 1930s they played a secondary role to private developers. The majority of suburban homes were built for owner-occupation and one of the features of the period was the decline in private rental compared to home ownership (Ball, 1983, 25–40). As house prices fell during the 1920s and 1930s, a standard semi-detached home in outer London costing between £400 and £600 came into the range of lower middle class and the upper stratum of skilled working class families (Weightman and Humphries, 1984, 114; Burnett, 1986, 255). Indeed, by 1938 nearly 20% of working class households across the country were owner occupiers (Ball, 1983, 36). Nevertheless, although local authority housing was a small component of urban change, it was significant in relation to suburban growth. The Addison Act of 1919 and Wheatley's Act of 1924 both provided subsidies for local authority housing. Between 1919 and 1929, LCC housing policy was oriented towards suburban cottage estates (Young and Garside, 1982, 153–64). The largest at

Becontree grew from 14 000 in 1924 to 103 000 by 1932, whilst others built at a similar time, such as at Bellingham, Roehampton and Watling, although smaller, nevertheless also introduced a substantial working class population to suburban localities (Young, 1934, 48 and 60).

In addition to improved transport and massive suburban growth, the spread of electricity was crucial to economic development on the periphery in two inter-related ways. First, it broke the locational dependence of industry on supplies of coal and helped reduce start-up costs. With the establishment of the Central Electricity Board in 1926 and the creation of the national grid, power costs for industry were considerably reduced. In London, rationalization of the numerous small and costly generating stations from the late 1920s lowered electricity costs still further. The outcome was that many new industries which relied on elec-tricity were attracted to the capital whilst in traditional trades, such as clothing, the availability of cheap power hastened the spread of machinery and helped create larger factories (British Parliamentary Papers, 1938, 118; British Parlia-mentary Papers, 1939–40, 315). Secondly, the expansion of domestic electricity created a large demand for household appliances that underpinned the new con-sumer industries. In 1935–36 South East England as a whole accounted for 43% of the total sale of electricity in the country (British Parliamentary Papers, 1938, 8–9). In 1918 only about 6% of homes were wired for electricity but by the late 1930s the majority of homes in outer London were connected, with the percentage rising in some areas such as West Ham and Hendon, to over 90% (Chisholm, 1938, 68–75; Weightman and Humphries, 1984, 129–30). Whilst almost all houses had electric lighting, a substantial proportion also had irons, vacuum cleaners, radios, fridges, fires and cookers.

Fuelled by improved transport, the availability of land for industrial devel-opment, adequate supplies of labour, and cheap electricity, the expansion of employment and factories was rapid. Between 1923 and 1937 the number of workers in London and the Home Counties covered by national insurance increased from over 2.4 million to nearly 3.5 million workers, more than double the rate of population growth over a similar period (British Parliamen-tary Papers, 1939–40, 296). Figures on factory construction during the 1930s confirm the vitality of the outer London economy. Between 1932 and 1938 1055 factories closed in London and 1573 were opened (the latter being 43% of the number opened in the country as a whole). This net increase was due entirely to growth in the outer ring. Thus, between 1934 and 1938 the LCC area suffered a loss of 191 factories whilst the rest of London gained a total of 429 new plants (Abercrombie, 1945, 39).

Industrial growth in outer London was primarily concentrated in two broad areas: the Lea Valley to the north-east and a north-west quadrant (Figure 2.4). Other concentrations of new industrial development existed but these were of lesser importance. In the lower Thameside region industries clustered around the docks or were strung out on lands adjoining the river. Industries dependent on large supplies of raw materials, such as food processing, sugar and oil refining, cement and paper mills, located in this area. Heavy engineering and vehicle manu-facture, including Ford's works at Dagenham, were also important. Elsewhere the only other new industrial area of note was along the Wandle Valley to the south where new engineering works producing electrical cables, motors, radios and refrigerators, together with food processing, oil and varnish works were located (Abercrombie, 1945, 42–43).

Compared to these minor clusters, industrial expansion in the Lea Valley and north west, was of much greater consequence (Figure 2.4). From the 1870s cheap

△ INDUSTRIAL AREAS ESTABLISHED BEFORE 1918
▲ INDUSTRIAL AREAS ESTABLISHED BETWEEN 1918 AND 1939

Source: Adapted from P. Abercrombie. Greater London Plan, 1944 (1945)

Figure 2.4   Industrial areas 1918–39

railway fares had attracted skilled workers north-eastwards into Tottenham, Edmonton and Walthamstow. In turn the availability of labour laid the basis for industrial expansion later in the century in and around the Lea Valley. From the early 1900s firms involved in traditional east London trades, such as clothing, metal and engineering, and furniture, began to locate in this area, attracted partly by labour but also by transport facilities and cheap land. From 1918, these industries were joined by others, including chemicals and electrical engineering works (Smith, 1933).

The second area of expansion lay to the north-west in a quadrant which stretched from Hendon in the north to Brentford in the west, with three main clusters of growth around Park Royal, Southall and along the Great West Road. New industries were of greater importance here than in the Lea Valley, particularly the fastest growing sectors of electrical engineering, vehicle and aircraft manufacture, and food processing (Table 2.7). Good transport links, including excellent access to railway facilities, canals and proximity to three major arterial roads, the North Circular, the Great Western Avenue and the Great West Road were major factors in development. From the late 1920s the opening of trading estates, such as at Park Royal, and the provision of new factory sites, added further incentives to expand production in this area.

The movement of factories to these new industrial areas derived from two sources: the decentralization of existing firms and the birth of new industries. From the turn of the century figures referring to the industrial development of the Lea Valley and West Middlesex suggest that outmigration of established companies and the establishment of new firms were of roughly equal impor-

Table 2.7 Origins of firms in Outer London 1900–32

Location	New firm		From London		From elsewhere		Total	
	Number	Percent	Number	Percent	Number	Percent	Number	Percent
Lea Valley	36	30.0	55	45.8	29	24.2	120	100.0
West Middlesex								
Hendon	22	33.9	28	43.1	15	23.0	65	100.0
Park Royal	37	43.0	39	45.4	10	11.6	86	100.0
Hayes and Southall	9	29.9	12	38.7	10	32.3	31	100.0
Chiswick, Brentford, Heston and Isleworth	23	43.4	14	26.4	16	30.2	53	100.0
Total	127	35.8	148	41.7	80	22.5	355	100.0

Source: Smith (1933)

Table 2.8   Causes of the location of factories in Greater London 1900–32

Cause	Lea Valley		West Middlesex		Total	
	Number	Per cent	Number	Per cent	Number	Per cent
Space requirements	29	22.1	30	10.1	59	13.8
Cheaper land, rents and rates	12	9.2	33	11.1	45	10.5
Provision of buildings	15	11.5	53	17.8	68	15.9
Transport facilities	16	12.2	35	11.7	51	11.9
Proximity to London	11	8.4	68	22.8	79	18.4
Cheap labour	11	8.4	5	1.7	16	3.7
Owner lived near factory	16	12.2	2	0.7	18	4.2
Others	21	16.0	72	24.2	93	21.7
Total	131	100.0	298	100.1	429	100.1

Source: Smith (1933, 55, 85)

tance. Some minor differences existed between both areas. In West Middlesex new firms were more significant than in the Lea Valley, particularly in the Park Royal and Chiswick–Brentford clusters. In contrast, outmigration of existing firms was more characteristic of the Lea Valley development (Table 2.7). Access to the London market, improved road, rail and canal links, a supply of both skilled and unskilled labour and the availability of land were important factors in each of these areas. However, as Table 2.8 suggests, different requirements were emphasized in each district. In West Middlesex access to the London market and the provision of suitable premises were of greater importance, whilst in the Lea Valley it was availability of space, labour and good transport links. As Hall (1962, 137) points out, these differences represented traditional distinctions within London's manufacturing trades, with access to the market being necessary for bespoke trades in the west, whilst cheap labour and transport facilities were of greater importance for the mass production trades of the east.

For the inter-war years Table 2.9 illustrates those factory trades which were expanding at the fastest rates. In absolute terms, vehicle manufacture was of paramount importance. Agglomeration economies and a high degree of interdependence existed between body makers, component manufacturers and car assembly works. Consequently, with the major exception of Ford at Dagenham and the manufacture of heavy vehicles in outer districts, such as Southall and Watford, the majority of car makers were located in western districts with notable concentrations in Willesden, Acton and Park Royal (Martin, 1966, 39–44). In terms of relative growth, pride of place went to the aircraft industry which more than doubled its workforce during the period. Aircraft makers in north-west London, around Cricklewood, Hendon and Park Royal, were attracted by the location of motor vehicle and component manufacturers, access to aerodromes and the availability of large amounts of space (Martin, 1966, 42–43). The expansion of the building supply industry reflected the tremendous growth in suburban construction, as too did the demand for household utensils and furniture. Development of the packaging industry was indicative of the growing importance of consumer durables and pre-packaged processed foods, whilst the expansion of management and administrative functions called for more printed stationery. Growth in each of these industries reflected the importance of the new consumer market and the impetus to economic development provided by the forces of urban expansion.

Table 2.9   High growth factory trades in Greater London 1924–30

Sector	1924 Numbers employed	1930 Numbers employed	Per cent increase
Aircraft	4895	9878	101.8
Building materials	5231	8737	67.0
Cardboard box	7741	11908	64.5
Manufactured stationery	13309	19478	46.4
Hardware, hollow ware, metallic furniture and sheet metal	13749	20051	45.8
Motor vehicle and cycle	37088	51853	39.8
Furniture and upholstery	29682	38277	29.0
Electrical engineering	62151	76554	23.2
Greater London	798395	884516	10.8

Source: Census of Production 1930, part V, general report, 148–9

Note: Figures refer to trades employing greater than 5000 people

Changes in the economic geography of manufacturing were accompanied by structural transformations relating to the scale, organization and methods of production. In London, where small-scale production had dominated manufacturing, growth in firm size was a noticeable feature of the inter-war period. Even allowing for differences in methods of compilation, Table 2.10 reveals the tendency towards larger-scale production. Whilst workshop production was still important in some traditional trades such as clothing and shoemaking, in newer trades, such as electrical engineering, cables and vehicle manufacture, large firms dominated. In printing, 24 firms employed more than 500 workers each, whilst in construction, noted during Victorian times for the proliferation of small units, larger firms were much in evidence (Llewellyn Smith, 1931, 54–55; Llewellyn Smith, 1933, 267). Moreover, growth in firm size was not confined to manufacturing and construction. Six of the 14 firms in London that employed more than 2000 workers were large department stores (Llewellyn Smith, 1933, 5 and 8).

Enlarged units of production were in turn associated with several other changes. In the 1920s powered factories for the first time came to outnumber handicraft based workshops in London. Larger units allowed for a greater subdivision of labour and the consequent spread of mechanization. Commenting in 1934 on the nature of change, Llewellyn Smith noted:

> In the last twenty years the manipulative industries have entered a new stage of the development of the technique of production. The changes include the introduction of semi-automatic and fully automatic machinery, which tends to eliminate the skilled worker; the installation of mechanical handling by means of travelling belts, tiering machines etc, which eliminates manual heaving and carrying, and the study of scientific layout and management to secure a high quantity and consistent quality of output. (Llewellyn Smith, 1933, 255–56)

This set of transformations represented a common thrust throughout the economy, encompassing trades as diverse as cigarette making and clothes washing, retailing and office work, but it was in the newer trades and in the larger factories that the pressures for change were the keenest (Hutt, 1937, 168–90).

Table 2.10   Size of employment units in London 1851 and Greater London 1930

Size	1851		1930	
	Number	Per cent	Number	Per cent
Less than 10	11807	86.0	10859	33.4
10–49	1705	12.4	16205	49.8
50–99	137	1.0	2934	9.0
100+	80	0.6	2537	7.8
Total	13729	100.0	32535	100.0

Source: Census 1851, Llewellyn Smith (1931, 472–5)

Note: Figures for 1851 refer to male employees only whilst those for 1930 refer to male and female employees

Within larger factories and workshops the subdivision of production and increase of repetitive and more specialized tasks laid the basis not just for mechanization but also for increases in the proportion of semi-skilled and unskilled assembly line workers, many of whom tended to be women and juveniles. Llewellyn Smith noted in relation to the metal trades that:

It is the mechanization of production, therefore, which is responsible both for the altered relation of the craftsman to other types of labour, and also, since the majority of female workers are process workers, for the altered proportion of male and female labour. . . (Llewellyn Smith, 1931, 133).

In electrical engineering and electric cables between 1923 and 1930 the proportion of women and juveniles employed rose from 27.8% and 42.5% respectively to 32.9% and 50.4% of the workforce (Llewellyn Smith, 1931, 2 and 205). In food preparation, mechanized production and handling of pre-packaged items increased considerably, particularly in relation to branded goods such as bread or soups, with similar results (Llewellyn Smith, 1933, 27 and 108). Following the First World War the tremendous rise in demand for tobacco hastened the spread of machinery that could turn out 50 000 cigarettes an hour, not only rendering hand-rolling obsolescent in all but cigar making but ushering in a far greater proportion of female workers.

Nor was the service sector immune from invasion by machines. In washing, small hand-based cottage laundries were replaced during the period by much larger powered laundries (Llewellyn Smith, 1933, 373). In white collar work the amalgamation of firms and growth in size was paralleled by the development of management staffs operating in larger office units. As the size of offices increased and the amount of paper multiplied so the need to improve internal communications and speed up processing of information led to the development of new forms of office technology. Internal phones, machines for calculating, copying, dictation, duplication and envelope sealing, not to mention the continued spread of typewriters, all appeared in greater profusion from the late nineteenth and early twentieth century (Llewellyn Smith, 1934c, 284; Hall and Preston, 1988, 74–8).

Many of these structural changes were linked to the establishment of factories and offices by large multinational firms, most of which were American. Although a few foreign firms had opened factories in London prior to the First World War, from the 1920s growing international competition, improved communications, and the threat of import tariffs led to a surge of foreign investment. Both before and after the First World War over half the initial manufacturing plants established by foreign companies were located in and around London. Between 1918 and 1944 70% of foreign factories and 65% of their employment were in the region, with US firms accounting for two-thirds of these (Jones, 1988). In electrical engineering, Westinghouse and General Electric, both US companies, dominated production, whilst Ford at Dagenham and General Motors at Luton occupied a similar position in vehicle manufacture. The imposition in 1931 of import tariffs led to a further rush of foreign investment, much of which was located in the suburban fringe. Indeed, colonization of new industrial suburbs was often pioneered by American companies. Factories belonging to American companies, such as Hoover, Gillette and Firestone, lined the Great West Road from the 1930s with similar development taking place along other new arterial routes. Heinz, whose plant in Southwark was opened in 1905, was one of the first to migrate to outer London when it moved its site westwards to the Park Royal estate in 1925 (Martin, 1966, 144).

In many instances foreign firms introduced far reaching changes to production and marketing. American companies pioneered the spread of scientific management, notably the Bedaux system which attempted to develop a strict scale of payment based on time and motion studies of production tasks. In total, during the late 1920s and 1930s over 250 firms in Britain adopted this system, including several large companies in London involved in food production, electrical cables and woodworking (Glading, 1934; Barratt Brown, 1934; Hutt, 1937, 170; Littler, 1983). Assembly line production at Ford revolutionized vehicle manufacture throughout the industry. In turn, large scale production required high volume marketing and this depended on the development of mass consumerism, fostered and to some extent created by advertising. Here, too, American companies pioneered new methods, using radio, newspapers and the cinema to spread the word about cars, creams and toothpaste.

Such changes in the scale and methods of production generated considerable pressure on traditional labour relations. However, in the country as a whole, apart from the General Strike in 1925, the period from 1922 to 1939 witnessed comparatively low levels of labour militancy, both in terms of the number of strikes and the total numbers involved (Cronin, 1979, 126–38). In keeping with the trend, and also given the propensity in London towards relatively peaceful industrial relations, the pattern in the capital was the same. Between 1921 and 1936 only 181 of the 886 large strikes in the United Kingdom occurred in London, and of these only 41 were in the new consumer industries (Daly and Atkinson, 1940). There were occasions, of course, when militancy punctured the tranquillity: union recognition was a major cause of disputes, notably during the 1930s in the aircraft and chemical industries; enforcing the closed shop engaged the efforts of several trades; the replacement of skilled workers by youths, trainees, women or semi-skilled operatives was often a source of contention (Branson and Heinemann, 1973, 128–29; Hinton, 1983, 153). However, such disputes were the exception rather than the rule. Within the workplace, the slow adaptation of unions to the needs of workers in the new industries, their inability to break free from a narrow emphasis on craft, company hostility towards unions, and the problems of organizing semi-skilled and unskilled labour, particularly women and juveniles, inhibited labour militancy.

Beyond the workplace, other impediments to labour militancy existed in the suburbs. Compared to inner districts, the suburbs were characterized by higher wages and lower unemployment, resulting in fewer households beneath the poverty line. Unemployment was lower both within the outer ring and on the new LCC estates. In part this reflected the distribution of new industries, which were less liable to downturn, but it was also because the estates themselves tended to have a relatively high proportion of skilled workers and a comparatively youthful age structure. Strikers who fell behind with their rents also faced eviction (Young, 1934, 25, 116 and 144). Higher wages and lower unemployment in turn resulted in lower levels of poverty. In Acton and Willesden, for example, the percentage of persons in poverty in 1929 was 2.8 and 6.0, respectively compared to 9.5% for the County of London (Llewellyn Smith, 1934b, 124 and 132). Even in eastern suburbs, such as Leyton and Walthamstow, where the figures were 6.9% and 7.3% respectively, poverty levels were lower than in the inner districts (Llewellyn Smith, 1932, 148). Relative affluence was associated with satisfaction over living conditions. Evidence from LCC estates at Becontree, Roehampton and Watling suggest that levels of tenant satisfaction were relatively high (Cronin, 1984, 85). But elsewhere similar expressions of satisfaction prevailed.

With the outer ring of London by the 1930s containing a greater proportion of the city's population than the old, inner core, the transformation of the economic and social geography of London was of far deeper significance than just a question of numbers. As George Orwell remarked at the time, it was in 'the new red cities of Greater London' that the old class distinctions were being broken down and an indeterminate stratum of technicians and higher-paid skilled workers was being fashioned that was in tastes, habit and manners closer to the middle class than to their more proletarian origins (Orwell, 1982, 68–69). For him:

> The place to look for the germs of the future England is in light-industry areas and along the arterial roads. In Slough, Dagenham, Barnet, Letchworth, Hayes – everywhere, indeed, on the outskirts of great towns – the old pattern is gradually changing into something new. In those vast new wildernesses of glass and brick the sharper distinctions of the older kind of town, with its slums and mansions, or of the country, with its manor-houses and squalid cottages, no longer exist. There are wide gradations of income, but it is the same kind of life that is being lived at different levels, in labour-saving flats or council houses, along the concrete roads and in the naked democracy of the swimming pools (Orwell, 1982, 68–69).

Conclusion

In many respects little had changed in the London economy over the century and a half covered here. The sectoral composition of the workforce remained stable. At the end of the period, as at the start, most Londoners derived their income from services rather than manufacturing. Indeed, the dependence on services goes some way to explaining the stability of the metropolitan economy during this period. In turn, the demand for consumer goods from the London market underpinned the wide range of manufacturing trades in the capital. Both factors help explain the continued vitality of London, even during periods of depression and regional decline elsewhere.

There was continuity, too, in relation to the location of activities. The centre maintained its hold on commercial and financial activities, although it was less successful in sustaining its manufacturing base. But although industrial production deserted the centre, the centrifugal movement followed a clear cut pattern with trades migrating outwards along sectors linked to the old Victorian manufacturing belt. As Hall has remarked, '. . . it does not therefore appear that the inter-war industrial growth of Middlesex saw any radical departure from the principles which have traditionally governed the structure and location of London industry' (Hall, 1962, 137).

To emphasize continuity, however, is not to deny the importance of change. In the nineteenth century pressures on the London economy resulted in restructuring and relocation of manufacturing trades. The subdivision of production and the concentration of industry in the eastern districts were important realignments in the local economy. Much more significant, however, were changes which took place from the 1900s, particularly in relation to demographic growth and the massive geographical expansion associated with large-scale suburbanization. London became greater and greater each decade and production expanded in tandem with urban growth. Indeed, the two were inextricably linked, for suburbanization not only provided the demand for new products, but also the workforce for industry and services. Rising levels of home ownership, coupled with the creation of a new infrastructure both inside and outside the home, helped to sustain high levels of demand for the new consumer goods and services which appeared during the inter-war years. Meanwhile new forms of production, involving assembly line work and greater mechanization, tapped into growing pools of semi- and unskilled female and juvenile labour located in the suburbs. So, whilst the old Victorian manufacturing belt decayed, the outer ring gained in economic vitality. New structures of work emerged based on larger units of production and involving higher levels of management control. Often the new methods of work were associated with US and other foreign companies. By the end of the period outer London had become a region of vigorous economic growth and part of the arena for the investment of international capital.

The new metropolis was at the epicentre of the processes of economic and social stabilization. For those in work – and the proportion rose over the period as a whole – wage levels were sufficiently high and employment sufficiently stable to sustain higher levels of consumption and to dampen any general expression of labour militancy. Indeed, if anything the tide was flowing strongly in the opposite direction. Mass consumption and better living conditions drew attention away from class conflict and towards compromise and accommodation. As Orwell so presciently noted at the time, London was the harbinger of a new cultural and economic order, a role which it has played in the past and one that it continues to perform in the present.

3

Change in Economic Structure and Opportunity

Martin E. Frost

Introduction

In spite of decades of employment and population loss London remains a labour market of enormous size and complexity within the British national system. Its changes, its problems and its physical extent are all on a larger scale than any comparable city in Britain and constitute a challenge to understanding the ways in which economic change affects the workings and structure of this labour market and, through this, the lifestyles and opportunities of people living in the city and beyond. Within this framework this chapter seeks to examine the impact that the growth of certain types of economic activity within London have had on the range and distribution of employment opportunities faced by its residents.

Through the later years of the 1970s and into the 1980s it has become clear that employment growth in British cities has been concentrated into a progressively narrowing range of service related activities (Marshall, 1988). The mix of private and public service employment growth seen in the early years of the 1970s has shrunk (Frost and Spence, 1984), as first public and then some private, services have shown signs of employment decline. It is now evident that in the early 1980s net employment growth in London was mainly limited to a restricted range of financial producer services surrounded by widespread employment losses across a broad range of manufacturing, public and private services. The growth that occurred tended to be concentrated spatially into central London (Frost and Spence, 1991). Such a situation places strain on a labour market whose primary function is to match together the characteristics of residents with the nature of the jobs which are available within the city. Clearly, in a period when the nature of jobs being created differs from the nature of those being lost, some kind of flexibility on the part of residents is needed for effective matching to take place.

In a simple neo-classical model of an urban labour market this flexibility takes the form of either mobility across occupational or skill boundaries or spatial mobility within the city system. The problems posed to residents in either case can be substantial. They were first widely publicized in the early inner area reports of the Department of the Environment (1977a, 1977b, 1977c) which, some 12 years ago, described the 'mis-match' between the skills and backgrounds of many inner-city residents and the characteristics of jobs available. This was seen to lead to long-term unemployment and, consequently, to social problems. Over the period since the publication of these reports the notion of 'mis-matching' has become a popular theme in many discussions of urban labour markets, generally identifying unemployment as the consequence of low levels of occupational and skill mobility.

In addition to these notable problems, other factors are growing in importance that have a negative influence on levels of flexibility and mobility in the labour market. The most general of these is the increasing participation of women in the labour force. In Greater London between 1981 and 1987 the proportion of all jobs held by women increased from 42.3% to 44.7%. This is in line with national figures of 42.6% and 45.7% respectively. This trend has a number of consequences. For one thing, women are more likely to be dependent on public transport than men. It is clear that, in single car owning households with two employed members, it is almost always the male earner who travels to work by car (Dasgupta, Frost and Spence, 1985). Additionally, women are more likely to face constraints on the location and working hours of a job due to family commitments. This effect can be seen in the distances travelled to work by women of various ages. In London, as in most British cities, profiles of work-travel distances are almost identical for males and females below the age of 24. After that the profiles diverge with female travel distances being approximately half those seen for men (Dasgupta, Frost and Spence, 1989). However, as this ratio declines only slightly for the older age groups, it appears that it is not solely family care that determines work-travel distances. Even so, it is clear that as a group female employees are more dependent on local work opportunities than men.

In addition to the changing male–female balance in London's labour market, more complicated constraints are increasing in importance. The growth of single parent families and families in which both parents need to work adds to the number of city residents who find it difficult to accommodate fixed and standard working hours. In addition, females from some minority groups may hesitate to travel long distances. The plight of people facing each of these constraints has been graphically discussed elsewhere (Greater London Council, 1985b; Huws, 1984; Brosnan and Wilkinson, 1987). Their importance for understanding the impact of economic change is that they add a further element of inflexibility and constraint to the process of matching residents to work opportunities within the capital.

These elements of inflexibility exacerbate the problems imposed on the labour market by the spatial concentration of growth referred to earlier. The more concentrated growth becomes, either in terms of particular types of work or in terms of particular areas of the city, the more difficult the matching process becomes and the greater the risk of unemployment or withdrawal from the labour force. The broader the base of employment growth with regard to opportunities for male and female workers, full and part-time work, service and non-service sectors, central and local locations, the more likely it will be that residents will find jobs that at least offer acceptable if not ideal possibilities.

In this context this chapter examines changes in the composition and distribution of London's employment through the early and middle years of the 1980s. It examines the nature of growth that has occurred in terms of its spatial distribution, its industrial characteristics, its male and female composition and the opportunities generated for full and part-time work. The objective is to assess the degree of concentration found in each of these and to assess the likely impact of this concentration on the range of job opportunities generated by economic growth in the capital.

The Characteristics of Employment Data

The data used in this Chapter are taken from the Censuses of Employment for 1981, 1984, and 1987. This data set represents the most comprehensive and up-to-date series for assessing employment change within areas of London. However, in common with most data sets it has characteristics which limit the freedom with which it can be used.

The first of these characteristics is inherent in most workplace based employment figures. These figures rely on the address of an employer as the critical factor in determining the allocation of employment totals to particular areas within a city. Minor changes of registered address or reorganizations of the way a company is structured can produce apparently significant changes in the distribution of employment that are not real changes of jobs 'on the ground'. Furthermore a company is classified industrially by its dominant activity. Small changes in the balance of what a company does can produce a disproportionately large shift from one category of activity to another.

In addition, the detailed practice of the 1981 survey differed slightly from the procedures used in later years. In 1981 the 'unclassified' category of employment was used where there was doubt over the industrial classification of an employer.

Table 3.1   Office areas within central, inner and outer London

Central London	Outer London
Borough	Acton
City of London	Barking
King's Cross	Barnet
Marylebone	Beckenham
Westminster	Bexley
	Brentford and Chiswick
	Bromley
	Burnt Oak
Inner London	Croydon
	Dagenham
Bermondsey	Ealing
Brixton	Enfield and Ponders End
Camberwell	Erith
Camden Town	Feltham
Clapham Junction	Finchley
Deptford	Golders Green
East Ham	Harrow
Fulham	Hayes
Hackney	Hornchurch
Hammersmith	Hounslow
Holloway	Ilford
Lewisham	Kingston
Plaistow	Orpington
Poplar	Romford
Shoreditch	Ruislip
Stepney	Sidcup
Stratford	Southall
Streatham and Tooting	Sutton
Tottenham	Uxbridge
Wood Green	Walthamstow
	Wembley
	Willesden
	Wimbledon
	Woolwich

More detailed data collection in later years allowed this practice to be discontinued. Secondly, within the data used in this chapter, no allowance was made in 1981 for employers not responding to the basic survey on which the employment counts are based. In later years some adjustment was made to compensate for this.

The net effect of these problems is that the employment data used here are more effective at indicating broad shifts in the structure and distribution of employment than in indicating detailed changes either in terms of industrial composition or the areas affected. For this reason the London system is divided into three broad areas. Central London is defined as the office areas which broadly make up the City of London and the West End together with a small fringe on the south bank of the Thames. Inner London surrounds this area, comprising the remnants of the Inner London Education Authority area (the old London County Council area) with the addition of an extension towards Haringey which is forced by some rather large office areas. Outer London is the remainder of the former Greater London Council area. The constituent Office Areas for these definitions are shown in Table 3.1. In terms of the industrial classification of employment data, a flexible approach is needed, combining categories from various levels of the 1981 Standard Industrial Classification in order to identify specific aspects of employment growth within broad service industry divisions. In general, small employment totals are avoided in an attempt to minimize the impact of irregular changes arising from the data characteristics discussed above.

Areal and Compositional Changes in Employment

The purpose of this section is to illustrate changes that have taken place in the composition of London's workforce through the 1980s together with the spatial evolution of its balance of employment.

The most noticeable characteristic of the figures contained in parts (a) and (b) of Table 3.2 is their diversity. Differences exist between sexes, between time periods, and between parts of London. It is clear that a significant element of the decline that has occurred between both 1981-84 and 1984-87 has been concentrated into male full-time employment. With the single exception of Inner London between

Table 3.2   Percentage compositional change in London's workforce

(a) 1984–1987	MFT	MPT	FFT	FPT	TOT
Central London	− 2.8	−11.7	7.7	−7.5	0.0
Inner London	0.1	4.4	13.4	−4.7	3.1
Outer London	− 2.2	12.3	5.0	2.6	1.1
(b) 1981–1984	MFT	MPT	FFT	FPT	TOT
Central London	− 0.3	10.8	5.3	9.9	2.6
Inner London	−10.4	5.5	− 6.2	−8.7	−8.3
Outer London	− 2.7	9.2	2.2	−1.5	−0.8

Source: National Online Manpower Information System

Note: MFT   Male full-time Employment
MPT   Male part-time Employment
FFT   Female full-time Employment
FPT   Female part-time Employment
TOT   Total Employment

1984 and 1987 this category of employment showed a decline in all areas for both time periods. The rate of change is quite variable but the general pattern is of decline even within the supposedly buoyant labour market of Central London. Decline was not, however, restricted to this category. The performance of female part-time employment showed significant aspects of decline. This was particularly marked in Inner London, where significant employment decline of this type occurred in both time periods. In the other two areas decline can be seen in at least one period, with the decline of −7.5% in Central London from 1984 to 1987 being particularly striking.

In contrast to these aspects of declining employment the remaining categories of female full-time and male part-time employment appear predominantly positive in their growth. In particular, both Central London and outer London showed sustained growth in female full-time working, while the rate of increase for this type of employment within Inner London after 1984 was the highest in the table.

In addition to variation in rates of change for categories of employment, it is also clear that employment growth and decline of areas of the city has not been a smooth and continuous process through the 1980s. In the earlier period from 1981 to 1984 net growth was concentrated in Central London with some decline, albeit, small for Outer London, taking place elsewhere. Through this period the central area increased its share of the city's employment at the expense of other areas. This process reverses in the later period, for between 1984 and 1987 the non-central areas showed employment growth while employment in Central London remained static in overall terms. The area of principal growth in the later period shifted to Inner London where particular improvements can be seen in the rates of change associated with both male and female full-time employment. This is an interesting change in the relative fortunes of areas of the city. Already, substantial investments in road and rail links into Central London are being discussed on the basis of forecasts associated with the apparent turn-around in the employment performance of the Central London economy in the early 1980s (Department of Transport, 1989b). It now appears with the evidence of these later figures that some of that dynamism identified in the central area is spreading outwards in the urban system reversing a long standing trend of overall employment loss from the Inner areas of the city.

These changes are interesting but, almost inevitably over only a six-year period, they represent marginal change within the overall character of London's employment structure. To shed some light on the basic structure on which these changes are taking place Table 3.3 shows the percentage of London's employment which falls in each category within the three broad areas of the capital. Comparison of the two parts of this table shows the relatively small changes in the shares of each part of the city. For the most part, changes are restricted to one or two percentage points. Of much greater significance are two other features revealed by these figures. The first of these is the importance of the individual areas as sources of employment within the overall London labour market. Only about one third of all jobs are found in the central area of the city. Commuters into Central London may hold the misconception that virtually all the employment of the capital is concentrated in the centre but this is far from the case. What is more, this situation has not been affected significantly by the much publicized growth of jobs in the central area over the last eight years. Indeed, rather more than two-thirds of all employment opportunities can be found in the remainder of London, with almost one half of all the jobs in the outer area of the city. Clearly the nature of job growth and decline outside Central London are of critical importance.

The second notable features of Table 3.3 is the degree to which the structure

Table 3.3   Percentage shares of employment for areas in London

(a) Shares in 1987

	MFT	MPT	FFT	FPT	TOT
Central London	34.8	25.3	39.6	17.9	33.5
Inner London	22.9	30.0	22.6	25.2	23.4
Outer London	42.3	44.7	37.7	56.9	43.0

(b) Shares in 1981

	MFT	MPT	FFT	FPT	TOT
Central London	33.9	28.9	38.2	17.1	32.5
Inner London	24.1	30.5	23.3	28.2	24.7
Outer London	42.0	40.7	38.5	54.7	42.3

Note: MFT   Male full-time employment
         MPT   Male part-time employment
         FFT   Female full-time employment
         FPT   Female part-time employment
         TOT   Total employment

of employment differs within the three areas, particularly with respect to female employment. It is clear that part-time female employment is substantially under-represented in the central area whilst this is partly balanced by its over-representation in the outer region. To illustrate these patterns of over- and under-representation, Table 3.4 shows the shares of employment categories expressed as ratios of the share of all employment in an area in 1987 (i.e. as location quotients). This shows that, while Central London has 104% of the number of male full-time jobs that would be expected from its share of all employment, it has only 53% of the number of female part-time jobs that would be expected on the same basis. Overall, these figures show the specialized nature of the Central London labour market with both categories of part-time employment relatively under-represented. In contrast, the outer area has about a third more part-time female employees than might be expected. What this implies in simple terms is that the job opportunities of the non-central area are particularly important for part-time female workers. At the same time, for part-time female employees, change and growth taking place in the centre are downgraded in importance by a relatively low rate of participation in that subdivision of the labour market.

But what does this mean in terms of the absolute numbers of jobs gained or lost by areas of the city over the period 1981 to 1987? Without doubt, job growth has been concentrated most strongly in female full-time employment. In marked contrast, male full-time employment shows equally consistent losses. The figures of Table 3.5 show that, over the six-year period, these losses amounted to more than 100 000 jobs. This slightly exceeded the gains made by female full-time employment over the same period. If figures for London's total employment

Table 3.4   Location quotients of employment categories, 1987

	MFT	MPT	FFT	FPT
Central London	1.04	0.76	1.18	0.53
Inner London	0.98	1.28	0.97	1.08
Outer London	0.98	1.04	0.88	1.32

Note: MFT   Male full-time
         MPT   Male part-time
         FFT   Female full-time
         FPT   Female part-time

Table 3.5   Employment change for areas of London 1981–87 (thousands of jobs)

	MFT	MPT	FFT	FPT	TOT
Central London	−20.2	− 0.8	50.5	1.4	30.9
Inner London	−47.4	3.9	14.4	−18.6	−47.6
Outer London	−38.4	11.6	27.8	3.0	3.9

Note: MFT   Male full-time employment
MPT   Male part-time employment
FFT   Female full-time employment
FPT   Female part-time employment
TOT   Total employment

performance are examined on their own, this substantial redistribution of job opportunities is concealed. The strong growth of female full-time employment also contrasts with the weak and mixed performance of the female part-time category. Here only small growth occurred in the central and outer areas with significant losses taking place in the inner area. Taken together, these figures mean that there has been an overall net loss of part-time female jobs. One important aspect of this is the very limited amount of growth taking place in Outer London, the area of the city that has the highest relative concentration of this employment type. It is clearly the case that, regardless of previous concentrations of employment, it has been female full-time employment that has been the most dynamic component of the London labour market through the 1980s.

In geographical terms, the consistent strength of growth in female full-time jobs has ensured a positive overall balance for Central London. However, it is interesting to note that, even prior to the stock market crash of October 1987, the central area lost more than 20 000 male jobs in the six years of the study. Decline can also be seen in the overall performance of Inner London, although this is almost wholly the product of a very poor performance in the early 1980s that was not quite offset by later growth. Meanwhile, the stock of jobs in the Outer London labour market remained more or less static but with a powerful internal redistribution of jobs away from male full-time employment towards other categories. The purpose of the next section is to consider the industrial character of these changes.

The Industrial Character of Employment Change

It is often difficult to identify clearly the nature of growth in the service sector when the industrial classifications found in the Standard Industrial Classification (1981 version) are very broad. Within a city like London, legal services have a significantly different distribution from advertising firms and probably show rather different characteristics of employment change over time. Yet it is only at the most detailed (Activity) level of the Standard Industrial Classification that these activities can be distinguished. At this level there are more than 300 categories which makes the compact presentation of results impossible. Thus, in this section, a selection of different levels of detail is used to present a general picture which uses broad categories that covers the whole workforce (the Division Level). This is followed by a more detailed description of selected service activities first at the Class Level and then at the Activity Level, in an attempt to identify the important bases of employment growth in London.

Table 3.6 Employment change by industry division (thousands of jobs)

(a) 1984–1987

	Central	Inner	Outer
Agriculture, forestry, fishing	0.0	0.0	0.0
Energy, water supply	0.8	– 1.6	– 4.3
Metals and minerals	– 5.7	– 1.2	– 6.0
Metal goods, engineering	– 4.8	– 9.7	–28.8
Other manufacturing	–29.9	3.6	– 5.1
Construction	– 0.1	0.0	– 1.5
Distribution, catering	– 3.1	– 1.8	13.0
Transport, communications	–17.2	–12.7	5.2
Banking, financial services	73.5	19.7	29.4
Other services	–12.6	28.7	14.4

(b) 1981–1984

	Central	Inner	Outer
Agriculture, forestry, fishing	– 0.1	0.0	0.0
Energy, water supply	2.0	– 5.4	– 1.1
Metals and minerals	– 1.7	– 5.6	– 4.9
Metal goods, engineering	– 3.7	–17.5	–37.3
Other manufacturing	0.4	–21.0	–14.9
Construction	– 4.4	– 8.7	– 6.5
Distribution, catering	6.3	– 6.6	16.5
Transport, communications	–14.5	– 8.5	– 4.9
Banking, financial services	31.2	5.2	36.3
Other services	16.1	3.3	6.5

Source: National Online Manpower Information System

The pivotal role played in the changing structure of employment by a narrow range of service industries can be seen clearly in the employment changes shown in Table 3.6. Nowhere is this more marked than in the case of the central area between 1984 and 1987. Over this period the substantial increase in jobs associated with banking, finance and other business services almost exactly offsets decline in virtually every other sector. This is an accentuated form of the pattern of change seen within this area in the earlier period. At this time growth in banking and financial services was the most prominent feature of change but this was complemented by growth in distribution, catering and the broad category of 'other services'. After 1984 this broader base of growth fell away to leave a sharply specialized pattern of employment increase in Central London. The net product of change over the two periods was more than 100 000 extra jobs in financial and business services with virtually no growth, and for some sectors significant decline, in the rest of the central area's economy. Furthermore, this is not a simple case of office-based jobs growing in a service industry to be set against non-office jobs declining in manufacturing. To some extent the decline in the traditional printing functions of Fleet Street can be seen in the decline of the 'other manufacturing' division but many of the other net job losses were losses of office-based jobs from non-service industry enterprises with offices in Central London. Unfortunately any accurate assessment of these effects will not be possible until after the results of the 1991 Population Census are available.

However, the lack of this evidence weakens only slightly the general conclusion, that we are seeing in Central London the rapid development of an increasingly specialized employment base with a traditional mix of activities being dominated by an expanding financial and business service sector. This is not to say that other employment opportunities do not exist. Even by 1987 there were still more than 300 000 jobs in the 'other services' division in the central area, with a further

197 000 in distribution and catering. Clearly the central area is far from being a single industry economy but, nevertheless, financial and business services commanded 38% of employment in 1987, and was growing, while all other sectors were declining. It will be interesting to observe if, and when, this proportion exceeds 50%. In the meantime it is clear that the most dynamic area of job growth in the capital through the mid-1980s was in financial services in Central London.

It would be wrong, however, to conclude that growth of financial services was a feature only of the central area. The figures of Table 3.6 also show significant increases in these activities in both Inner and Outer London. In particular, there is consistent growth over both time periods in Outer London (producing nearly 66 000 extra jobs). In the earlier period the rate of growth in Outer London was nearly 28% compared with a rate of just under 10% in Central London. Even between 1984 and 1987 growth in the outer areas was 18% while in the central areas it was only 20%. Thus, while it may be clear that financial services are increasingly dominant in Central London this is not because this area is claiming a disproportionate share of all growth in these activities. Strong growth can be seen in all parts of the capital but, in common with the central area, this growth in financial services is set alongside decline in many other employment sectors.

In the outer area the division between growth and decline is clear in both periods. It is essentially decline in manufacturing and construction set against growth in the service industries, with transport and communications employment sitting somewhere between the two (showing decline in the first period and growth in the second). Taking the two periods together, financial service employment is clearly the strongest focus of growth but increases in distribution and catering are also significant. In this latter division the rate of increase is fairly modest at 5% and 4% for the two periods but the employment base is sufficiently large to produce a significant number of extra jobs from these rates of increase. A similar argument applies to the 'other services' division where a growth rate in the second period of just over 3% produces an increase of more than 14 000 jobs. In overall terms this pattern of growth is far less specialized than that seen in Central London, and carries with it a broader range of employment opportunities across a range of service activities.

Industrial change in Inner London has characteristics that fall between the positions of Central and Outer London. In the period between 1981 and 1984 it shows a profile of general decline with only financial services showing any growth at all. This changes dramatically in the second period but, even then, growth is dominated by the combination of financial and 'other services'. It is a pattern that is not quite as extreme as that of Central London but one which certainly lacks some of the breadth of service industry growth seen in Outer London.

Description at the broad Division Level can frequently conceal important aspects of employment specialization within the service industries. The variable nature of employment growth and decline in the service sector can be seen clearly in the figures of Table 3.7, which show in more detail changes taking place in a selection of important service activities. Within the apparently buoyant areas of distribution, financial and other services there are significant areas of employment decline. Within Central London, wholesale distribution consistently lost jobs over both time periods, while employment in insurance also declined. Public sector services present a very mixed picture of limited growth interspersed with significant losses, while the apparent buoyancy of employment in hotels and catering diminished significantly in the second period.

Set against this is the consistent increase of business services, showing substantial growth in all areas over both time periods. This is supported in the later

Table 3.7   Percentage employment change within selected classes

(a) 1984–1987

	Central	Inner	Outer
Wholesale distribution	−16.2	− 4.3	2.3
Retail distribution	6.4	3.8	4.0
Hotels and catering	0.2	− 7.3	9.8
Banking and finance	17.5	28.0	12.9
Insurance	− 0.6	9.5	1.6
Business services	24.9	33.1	27.0
Public administration	−5.7	− 3.2	0.5
Education	− 1.5	2.5	0.3
Medical services	− 1.1	5.9	0.6
Other services	11.1	29.1	6.6

(b) 1981–1984

	Central	Inner	Outer
Wholesale distribution	−19.0	− 1.4	2.1
Retail distribution	5.4	− 7.2	3.8
Hotels and catering	11.9	− 5.0	15.7
Banking and finance	5.3	− 6.3	9.2
Insurance	−12.6	− 8.3	8.5
Business services	15.0	13.7	38.9
Public administration	8.3	5.1	− 2.8
Education	0.2	− 2.8	− 2.7
Medical services	− 9.7	−14.3	0.7
Other services	5.7	16.5	15.7

period by employment in banking and finance which showed a marked increase in its rate of growth. Apart from these categories only employment in the general 'other services' class showed consistent growth throughout the city. In terms of growth this is a picture of considerable, but not total, concentration in a narrow range of banking, finance and mixed business service employment. To investigate in more detail the precise nature of growth within this area of employment, Table 3.8 shows business service employment broken down to the most detailed level of the Standard Industrial Classification (the Activity Level).

With one or two clear exceptions, the striking feature of Table 3.8 is the rapid rate of growth to be found across a range of business service activities. The growth is not restricted to the narrowly defined financial area but extends into estate agents, legal services, computer services and particularly into the poorly defined business service group. Essentially this category contains the broad sweep of sub-contracting and consultancy firms that cannot be classified elsewhere. The central area in the years between 1984 and 1987 not only saw firms in this sector show a 50% increase in their employment in only three years but the number of jobs this represents makes up more than 40% of the total increase for this class, amounting to more than 20 000 jobs. Similar rates of increase can be seen in the other areas of the capital. In the outer area this limited range of activities contributed nearly 70% of all jobs gained in the more general business service class between 1984 and 1987.

The results all add to the impression of growth of employment in many areas of the city being concentrated into a narrow range of business service related activities. It is clear, however, that the nature of the jobs available in these activities need not be as restricted as their apparent industrial base. Thus, clerical skills or computing skills have considerable mobility between sectors. One feature of specialization that is particularly striking, though, is the degree to which

Table 3.8   Percentage employment change within business services

(a) 1984–1987

	Central	Inner	Outer
Auxiliary to banking	44.9	145.5	92.6
Auxiliary to insurance	1.2	4.3	−16.1
Estate agents	20.9	22.2	13.7
Legal services	26.8	27.1	13.0
Accountants	28.0	4.4	9.9
Prof., tech. n.e.s.*	24.8	40.6	6.1
Advertising	6.0	12.3	5.2
Computer services	23.9	48.9	34.4
Business services	53.1	51.2	66.4
Central offices n.e.s.*	−44.5	−37.0	−3.8

(b) 1981–1984

	Central	Inner	Outer
Auxiliary to banking	62.3	13.4	52.3
Auxiliary to insurance	5.3	80.4	31.4
Estate agents	25.1	37.9	34.4
Legal services	16.3	4.6	34.2
Accountants	6.8	−7.1	29.2
Prof., tech. n.e.s.*	7.2	−18.6	32.8
Advertising	4.4	7.9	111.3
Computer services	38.8	49.5	27.3
Business services	46.1	14.4	50.4
Central offices n.e.s.*	−34.7	61.2	28.3

Note: * These are activities that are not elsewhere specified. Many aspects of professional and technical or central office employment will be included in the industrial category appropriate to the company they are associated with.

Table 3.9   Full-time proportion of all employment growth, 1984–87

	Central	Inner	Outer
Banking and finance	100.5	106.3	103.4
Business services	95.0	100.0	79.1
Other services	94.2	78.3	79.3

growth in many of these activities is concentrated into full-time employment. To demonstrate this concentration Table 3.9 sets out, at the class level, full-time employment growth for selected classes expressed as a proportion of all employment change (in those classes). Within these figures a percentage greater than 100 shows that part-time employment actually declined in these classes over the period. The immediate conclusion is that growth is dominated by full-time employment in all of these categories with only a few values dropping below 80% in any area of the city. This concentration of growth in full-time jobs offsets to some extent the possible flexibility to be gained from transferability of skills between some of these service industries. However, the limitation of opportunity is concentrated on a category of employment, part-time working, rather than on any specific industrial grouping.

Thus, the net conclusions from this section point to a structure of economic change that has generated substantial employment growth within the London system over the period of the study. This growth has not, particularly in the second half of that period, been restricted to Central London alone. It has been spread widely across all areas but has been restricted to a limited range of predominantly financial and business service activities. In some areas, most notably Central London between 1984 and 1987, growth in these industrial groups effectively offset decline in other types of activity. However, within this specialized structure of growth there is a striking concentration in full-time rather than part-time employment. This implies that there has been a reduction in the range of job types available within the city. The next section considers some of these changes in relation to national employment change. The objective is to examine the degree to which London is simply following national trends or is developing along an individual path.

Employment Change in a National Context

The nature of London's employment change deviates in some significant areas from the national pattern. Table 3.10 shows the broad structure of national employment change for both periods of the study. Two features stand out in the comparison of London's employment performance with that of the nation. The first of these is the failure of employment in Central London to show improvement in performance in the later period that is comparable with the 2% increase in the country as a whole. Rather, employment in the central area was essentially static over this period (Table 3.2). The second feature is the strong growth of part-time employment in the country as a whole that is not reflected in the performance of any of the London areas. Clearly there are grounds for believing that the evolution of employment opportunities in London is progressing along a path that is significantly different from the rest of the country.

Table 3.10 Percentage composition of national employment change

	MFT	MPT	FFT	FPT	TOT
1984–1987	−2.2	14.0	5.7	6.8	2.0
1981–1984	−4.2	10.2	0.6	3.4	−1.2

Source: National Online Manpower Information System
Note: MFT Male full-time employment
MPT Male part-time employment
FFT Female full-time employment
FPT Female part-time employment
TOT Total employment

One important factor determining such a difference is the economic structure of the capital. To make allowance for this a shift/share analysis is used here which allows the rates of change of London's industries to be compared with their national counterparts. The proportionality shift in this analysis shows the amount (in numbers of jobs) by which the employment of an area would be expected to grow faster or slower than the national rate, as a result of the industrial composition of that area. The differential shift shows the degree to which the industries of the area grew faster or slower than their national counterparts.

A strong positive proportionality value for male full-time employment in Central London for the 1984–87 period in Table 3.11 shows that the area started the period with a favourable endowment of industries and would have bettered the national rate of overall employment change by some 47 000 jobs if these industries had only maintained rates of growth comparable with their national

Table 3.11   Shift/share analysis of London's employment (thousands of jobs)

(a) 1984–1987

	Central	Inner	Outer
Proportionality Shift			
MFT	46.9	10.5	15.2
MPT	− 2.4	0.6	− 0.3
FFT	25.6	4.6	4.6
FPT	0.9	1.2	1.1
TOT	67.0	17.9	21.3
Differential Shift			
MFT	−50.5	− 0.6	−14.4
MPT	− 7.9	− 4.5	− 0.6
FFT	−17.7	11.8	− 7.4
FPT	−14.4	−16.2	−12.6
TOT	−90.6	− 9.2	−35.5

(b) 1981–1984

	Central	Inner	Outer
Proportionality Shift			
MFT	46.7	16.1	10.5
MPT	2.6	0.3	0.1
FFT	21.6	4.6	2.1
FPT	4.2	1.5	− 0.9
TOT	72.0	24.0	13.5
Differential Shift			
MFT	−21.4	−44.7	1.6
MPT	− 2.4	− 2.1	0.6
FFT	− 4.2	−20.4	4.0
FPT	1.3	−18.8	−12.5
TOT	−28.2	−86.5	− 8.2

Note: MFT   Male full-time employment
MPT   Male part-time employment
FFT   Female full-time employment
FPT   Female part-time employment
TOT   Total employment

counterparts. In fact, the employment of Central London's industries failed to keep pace with comparable national rates of change to the tune of more than 50 000 jobs, producing a negative figure in the differential shift. In simple terms this means that Central London would have contained more than 50 000 more male full-time jobs if the employment performance of its industries had matched comparable national rates of change.

Extending description of the differential shift to cover all employment, it can be seen that the central area would have contained a substantial 90 000 extra jobs if its industries had simply grown or declined at national rates of change. This shortfall can be added to the equivalent deficit of nearly 30 000 jobs seen in the earlier period. Of all the categories of employment over both time periods only part-time female employment from 1981–84 shows a performance that keeps up with national rates of change. The critical question is whether these are signs of economic strength or weakness.

To determine this, one must think more broadly about the nature of Central London and its economy. Above all else it is an economy that faces constraints: physical constraints in the buildings that are available and economic constraints in labour supply and the ability of the workforce to travel to the centre of the city. In this setting, the figures of Table 3.6 showed that more than 100 000 extra jobs in banking, finance and business services had to be accommodated between 1981 and 1987. If the area had followed national rates of growth this figure would have been about 19 000 greater. This demonstrates the enormous strength of growth in these activities nationally but does not reduce the pressure of housing and staffing brought on by an addition of approximately 10% to Central London's total employment in 1981 within six years. In this light, the performance of Central London can be seen not as a process of growth or decline but as one of specialization. It is notable in this context that only banking and finance, business services and retailing grew significantly faster than their national counterparts in Central London after 1984; almost every other form of employment showed a relative decline. The bases of a new, more specialized Central London are becoming clear.

The relationship between Central London and the rest of the city is inevitably a complicated one. The rest of London is adjacent to the centre with less immediate pressure of physical constraints. But it too has problems of site availability and accessibility, while firms have to compete in a labour market that, if not the same as that for Central London, certainly has significant overlaps. It would, therefore, be naive to expect relative employment losses caused by physical and labour market constraints in Central London to be reflected by compensating gains in the rest of the city. Even so, Inner London significantly improved its employment performance between 1984 and 1987 even when this performance is set in a national context. This compares with 1981–84 when it suffered a comparative loss of nearly 90 000 jobs. After 1984 this comparative loss shrank to only 9000 jobs, with female full-time employment growing significantly faster than national rates would predict. This reversal of relative decline was built on some significant job increases. During the three years after 1984 full-time employment in banking and finance, and in business services, grew by 9000 jobs more than would be predicted from an already rapidly growing national rates of change. Also prominent in the inner area was the relative gain of employment in paper product manufacturing which includes the printing and publishing industry. One component of this was the movement of newspaper production away from its traditional centre in Fleet Street (which is included in Central London). It is interesting to note that the relative gain to the inner area was just over 8000 more jobs than would be predicted by national rates of change. The relative loss seen in the central area compared to the same national rate of change is just over 20 000.

A note of caution is needed, however, to point out that not all gains in the inner area can be attributed to private sector services or other private activities. One of the largest single relative gains of employment within the shift/share figures is contributed by full-time employment in education. This sector gained almost 11 000 full-time jobs in relation to the national rate of employment change. However, Inner London also lost over 13 000 part-time jobs by a similar comparison. Clearly the staffing policies of the Inner London Education Authority are critical here. These figures alone play a substantial role in determining the strong relative performance of female full-time work in the inner area and the weak relative performance of female part-time work in Table 3.11. However, the balance of the evidence supports a general conclusion that the position of Inner London improved dramatically between the two periods and that a significant part of this improvement arose from rapid growth in nationally expanding service sector activities.

The improving conditions of the inner area are not reflected in Outer London. Once industrial structure is accounted for, the performance of the outer area is seen to deteriorate from the earlier to the later period. Between 1981 and 1984 employment change in the outer area almost matched national trends, falling short only in the female part-time category. In the later period, a relative loss can be seen in all categories. The area ended the period with an overall relative loss of more than 35 000 jobs. Within this poor performance, few activities showed a positive picture relative to the national situation. The ubiquitous business services group showed a relative gain of nearly 5000 jobs, but many other services were less buoyant. Although the raw figures of Table 3.6 showed a 1984–87 increase of more than 14 000 jobs in the 'other services' category this was achieved with many activities growing at less than the national rate. Apart from business services it was only in retailing and in hotels and catering that employment levels outstripped their national counterparts, and here, only in full-time employment.

Thus, in broad terms, the three areas of London present three rather different profiles of economic change. Central London seems to be characterized by increasing constraints limiting its growth as national growth pressures over a wide range of business services are focused on a limited area and workforce. In contrast, Inner London appears to be increasing in dynamism within the service sector even after the (perhaps temporary) effects of the now disbanded Inner London Education Authority are discounted. Against this the employment performance of Outer London is discouraging. Over most of its industrial structure, activities have not matched national employment changes, particularly after 1984. In this later period only a few aspects of the service sector have shown growth at above national rates. It is revealing to note that, within business services, the category of employment showing the fastest relative growth is the part-time female category, which is in marked contrast to trends in all other parts of the city.

Conclusion

One clear conclusion to emerge from the pattern of London's employment changes is that there has been considerable concentration in the growth of full-time rather than part-time work. Central London has always tended to be an area with a marked degree of specialization in full-time work (Frost and Spence, 1984). Although the causes of this dominance of full-time work have never been properly investigated, it is plausible to relate it to a combination of the specialized types of activity found in Central London, linked to the time and effort needed to travel even short distances to work in the centre. It seems, however, that the strong channelling of growth into the finance and business service sectors is spreading the tendency towards full-time work more broadly across London. This is in contrast to the national pattern of change that shows much stronger part-time growth, particularly after 1984. In labour market terms this difference is open to at least two different interpretations. It could be argued that, in the absence of sufficient demand for full-time employees outside London, a greater proportion of work seekers are taking part-time employment when they would prefer a full-time position. Alternatively, it could be that residents seeking part-time work in London are being disappointed by a dominant demand for only full-time employees. From aggregate employment figures it is impossible to evaluate these alternatives. However, if this trend continues it will produce a labour market in London that offers fewer and fewer opportunities for residents who genuinely need or prefer part-time work. They will be forced increasingly into those activities, such as retail distribution in large stores and supermarkets, that have become substantial sources of shift based part-time jobs (Greater London Council, 1985b).

A second feature of London's employment change is the substantial concentration of growth into a narrow range of producer services. Even within this restricted group there is further concentration into banking and finance and the broader category of business services. There are also some notable absentees from the list of expanding activities. It is clear that employment in insurance expanded relatively little in the London area during the early to mid-1980s. The rate of expansion in advertising was also modest. By contrast, the rate of growth of the residual business services activity was remarkable. In Central London those activities based around management consultants, market research and public relations consultants increased their employment levels by almost a half between 1981 and 1984 and then by half again after 1984. In Outer London the rate of growth was, if anything, slightly faster, with Inner London attaining a comparable growth rate in the second period. Together with services specifically linked to banking, these are the fastest growing activities in the London system. They represent the peak of a limited range of private sector service-related activities that are the sole basis for significant employment growth throughout the London system.

Such a high level of dependence on a limited range of service activities inevitably creates a fear that any problems in the financial sector might have severe employment effects and also that the narrow range of activities might further restrict the range of job opportunities available within the city. It is relevant to note that the 1987 Census of Employment was conducted immediately before the stock market crash of October 1987. Over the last two years wildly different estimates and forecasts of job losses associated with the subsequent downturn in levels of stock market activity have been made. To set against this apparent exposure it is interesting to observe that the fastest rates of employment growth, and some of the largest absolute increases, have come from the broadly based business service area (including research consultancies and market research organizations) rather than from the narrowly defined financial sector. Clearly the possible linkage between these activities cannot be established from employment figures but the evidence of employment change suggests that there might be a broader base to growth in London than the well publicized expansions in banking and finance alone suggest.

However, in industrial terms, the range of employment growth is still limited. This clearly has some implications for labour market opportunities in the capital. Throughout the 1980s the grip of office-based employment has strengthened in most parts of the city. However, the effect that this has on the labour market depends on the level of interchangeability of skills between static or declining activities and more dynamic service sectors. There is anecdotal evidence to suggest that the level of demand for secretarial skills remains high. Case study evidence suggests that, in the country as a whole, secretarial and other support staff make up at least one quarter of the employees of some large accountancy firms (Marshall, 1988). At this rate, a significant part of the employment increases described above will be in secretarial employment. Furthermore, these secretarial skills will be transferable between activities. What is not clear is what the equivalent labour market position is for residents involved in the static or declining areas of public sector services. Relatively few of these are likely to have skills that are easily transferred into private sector business service employment (a position they share with many of those who are still losing jobs in the continuing decline of manufacturing in the city). The true levels of labour market stress caused by these changes cannot be assessed adequately from aggregate employment statistics. It is an issue that is critically in need of more detailed survey-based research.

A final set of conclusions relate to the spatial character of employment change. There are clear signs that, particularly in the later period, Central London was increasingly experiencing constraints in its growth and that the employment growth generated by its activities was dropping further behind national rates of change. There is slender evidence to suggest that Inner London derived some benefit from this, showing a positive performance in the second period of the study which was quite different from its recent history. Outer London showed the opposite trend with an employment performance that worsened in the second period. Many interpretations are possible from these results. It seems reasonable to conclude that we are seeing the emergence of a more specialized core of the city, based around finance, business services, and to some extent, retailing. This is essentially a 'filtered' product of what already exists in Central London. However, it may be that the traditional core is expanding to include parts of Inner London. The most prominent example here is the Docklands area where massive office development promises to create a third office-based employment 'node' to add to Westminster and the City of London. Early evidence of this is possibly apparent in the later period of the study. This leaves the problem of Outer London. One explanation of its performance might be that, faced with spiralling costs in Central London, firms in private sector services have decided not to relocate activities to other parts of the capital but to migrate to more distant centres. There is recent evidence that supports this (Leyshon and Thrift, 1989). Another explanation might be that the strength of demand for labour in the central area drains the effective labour supply of the outer area thus inhibiting its growth. The problem with this argument is that growth in Central London has been predominantly redistributive with little overall net increase in its share of all employment in the city. On balance, it seems more likely that, after 1984, Outer London was bypassed by many (though not all) areas of employment growth. That it achieved the highest rate of part-time female employment growth of any area for business services is evidence of the detailed 'sifting' processes of locational choice that are evident in the evolving structure of economic activity in the London labour market.

It is clear that the economic structure of London is undergoing rapid change. Strongly focused national change, joined with the increased importance of international financial trading, have produced a pace of change of greater magnitude than has been seen for many years. The changes produced by these forces are significantly affecting the range, type and location of job opportunities which are open to residents of London. The direction and extent of these changes in the future depends crucially on both the performance of our domestic economy and the success of London as an international trading and financial centre in the face of developing international (and particularly European) competition.

4

A New Geography of London's Manufacturing

F. E. Ian Hamilton

'. . . we can hardly foresee any causes, apart from the decay of Britain itself, which
shall lead to a failure . . . of London. Rather must we consider whether, with more
general distribution of industrial activity over all lands, the function of London as
chief clearing-house for . . . the world's exchanges be lost or retained or magnified'.
(H. Mackinder, 1902, 339-40)

The world economy has undergone gradual, then accelerating, tendencies towards
'globalization' in the past 40 years. Economists, geographers, planners and politicians, for varied reasons, were slow – even reluctant – to recognize this fundamental change. Moods and opportunities in world economic, military and
political arenas in 1990 point to further intensification of globalization beyond
2000. The nation state's rise to power as the prime force shaping the economies
of any place – city, region or locality – peaked generally between 1960 and 1990.
Decisions implemented by firms, like multinational enterprises, and transnational
institutions or organizations, such as the European Community (EC) or General
Agreement of Tariffs and Trade (GATT), have progressively diluted, substituted
or replaced central government influence, protection and Keynesian demand-side
policies (e.g. public spending) by cross-border, international integration. This has
exposed places to sharper global competition and a need to adjust to supply-side
economics (i.e. cost consciousness). These forces have gradually tied segments
– if not the whole – of regional and urban economies into a complex web
of global operations. Places, however, function through people and organizations with potential to shape these processes actively, but they may also suffer
developmental weaknesses, inducing passive dependency on places, people and
organizations elsewhere.

A World Economic Centre

London is such a place, but it is no ordinary place. For four centuries, it has
conducted trade and transactions on a world scale. The largest city in the
world a century ago, it now ranks 20th in population, having been overtaken
by the metropolitan foci of states which displaced the UK as a leading global
economy (like New York, Los Angeles, Chicago, Tokyo, Paris or Moscow),
and by primary centres like Mexico City, São Paulo, Seoul, Bombay, Beijing or
Jakarta in demographically-exploding, or newly-industrializing countries (NICs).
This suggests that London is in relative decline. But most of these cities grew

in an uncontrolled fashion. They have not been subject to the four decades of stringent planning regulations which have contained Greater London's sprawl within a protected green belt. Nor have they experienced policies to assist relocation of existing, and diversion of new, industry, administration and research to places beyond a green belt. After 1948, when building the new towns began, these processes intensified linkages between the core (Greater London) and a broadening area (first, the outer metropolitan zone, later also the outer South East) forming a 'dispersed metropolis' (Chapter 9), and forging a complex, integrated *city region*.[1] On this scale, London retains much of its former international importance.

The city's economy has undergone periodic metamorphoses against a backcloth of continual change: internal restructuring has accompanied, responded to, and created, dynamic external relationships. Aspects of such internal change are outlined by Frost regarding employment change and Diamond on the growth of financial services in Chapter 3 and Chapter 5 respectively. Less than two generations ago, London was the world's busiest seaport and, except for New York, the biggest manufacturing centre. Today, manufacturing and port appear as mere shadows. Yet, through Tilbury, London still ranks in the top 20 seaports of the non-communist world, though much of its traffic is transposed to ports in the South East like Dover and Felixstowe with roll-on/roll-off or container handling facilities. Manufacturing retains global significance, employing some 450 000 workers in Greater London and nearly one and a half million in the city region. The metropolis has become a formidable world 'control point', a headquarters city for national and foreign businesses, while it jockeys with New York and Tokyo for leadership in financial transactions. Connected with this growth, it has also attracted a major share of the soaring volume of world air traffic: Heathrow and Gatwick handle more passengers annually than the airports of any other city, save New York and Chicago. It is part of a global labour market and a cosmopolitan tourist centre. Though not comparable with New York in numbers or proportions of immigrants, London has drawn thousands from the Commonwealth and from Europe; and, for every three inhabitants, four foreigners visit it annually.

Indicative of Greater London's economic size is its Gross Regional Product. This amounted to US$78 billion in 1985 (Central Statistical Office, 1989), equal to 17.1% of the UK GDP and 3.9% of the European Community GDP – making the city the 25th largest economy amongst the world's 212 independent territories or nations. London created as much GDP as all the last 108 smaller and least-developed of those 212 states *combined*. And its GDP exceeded the sales turnover of all multinationals, excepting only General Motors and Exxon.

London in the International Production System

The city economy is a complex production system. It comprises a *population of organizations*, including private-sector manufacturing and service firms and government ministries, which act as producers and consumers; and a *regional* economic and social environment also shaping the demand and supply sides of London's economy. These operate within a *hierarchy of production systems and environments* at the UK, European Community, capitalist-world and global levels. Each level generates diverse 'impulses' in varying strengths – regulations, policies, competition, demand, financial trends or new technology – to which the London population of firms may need to respond directly to sustain their growth or survival. Such external forces may also combine with the impacts of changes in the population of organizations or in their types of activity to alter the real or perceived value of elements of London's regional conditions — with further indirect effects on local populations of firms.

London's production system is thus open to various regional, national and international linkages, contacts or influences, while simultaneously exhibiting strong internal cohesion and integration. For example, the city's emergence as a banking and insurance centre in the 18th and 19th centuries was rooted in the need to fund and cover a growing volume and array of cargoes on long, hazardous journeys across the seas to British ports. The seaport, commodity trading and financial services became interdependent. Indeed, the economy diversified so much that many distinctive production sub-systems are iden- tifiable which, in greater or lesser measure, have retained their identity in London's landscape over long periods. The large scale of operations needed to serve London, UK and foreign markets, and the scope for agglomeration economies, allowed each sub-system, or population of organizations, to achieve a critical mass and occupy extensive, specialized 'quarters'. Regular face-to-face contacts between such populations of firms buying, selling, designing, manufac- turing or sub-contracting gave strong cohesion. Less frequent contact among firms in different, but related, sub-systems permitted their spatial separation, while maintaining close proximity.

Four major types of productive sub-systems are distinguishable. All evolved to exploit London's critical, and apparently timeless, comparative advantage in the international chain of production: as an entrepot for flows between global places and a gateway to regional or UK national markets. The four types are as follows:

A materials-based sub-system of cargo-handling and manufacturing which links materials imports (from abroad or other UK regions) and sales of manufactures to the city's, other UK and foreign markets. In the past, this sub-system had several distinctive parts. Food and drink firms sited in the east near the port, or in the west along the Grand Union Canal, like Tate and Lyle (sugar), Sarsons (sauces), Gordons (gin) or Peek Frean (biscuits), have attracted other manufacturers to supply packaging (Metal Box) or bottles (Key Glass), printed packets or advertising. Furniture makers clustered in north London. Chemicals, oil-refining, paper-making and power stations were down river to consume bulk imports of oil, timber, North Sea gas and sea-coal. Jewellers, fashioning precious metals or stones, and linked clock, watch, optical or precision firms developed in Clerkenwell, clothing in the East End, printing in and near central London, with engineering in suburban London (Martin, 1966).

A transactional sub-system which includes several distinct populations of City firms in finance and commodity trading which conduct global business. For example, clustered around the Bank of England are British and some 450 foreign banks, like the National Westminster, Banco de Bilbao y Vizcaya or the Bank of Hiroshima. Associated activities include the London Stock Exchange and linked stockbrokers, like Warburg or Lazard Brothers, insurers centred on Lloyd's and overflowing west into Holborn, like Prudential, Commercial Union or Royal, commodity exchanges and futures markets for items like gold, tin, oil, cocoa or ivory, and accountants who engage in various functions which, like consultancy, overlap with the next sub-system.

An information sub-system which focuses on data and information control, decision-making and the media. London's evolution into an imperial economic,

political and cultural capital and, in the latter part of the 19th century, as a centre of a democratizing Britain, fostered a boom in printing and publishing news-papers, journals and books. Fleet Street became the hub for firms like Reuters, so making it pre-eminent in international news. Innovations such as radio and television led to a diffusion of media organizations like the British Broadcasting Corporation (BBC) or Independent Television, heralding a new era in world information in which English became *the* global language. In addition, in the West End, there are concentrations of firms specializing in company advertising and films which are partly linked to the last sub-system.

An administrative sub-system which comprises government, business, law and professional bodies. They occupy distinctive nodal spaces – central government in Westminster, lawyers at Temple, company headquarters and offices in a zone enclosing the legal-political core from the City of London along Euston Road through the West End to Knightsbridge and Victoria, and along the South Bank from Vauxhall to Southwark. This sub-system is complex, for it embraces both the fully-integrated (internalized) offices of government ministries like the Department of Environment or firms like British Petroleum (BP) and British American Tobacco (BAT), and externalized specialist office-servicing firms like employment bureaux (placing temporary staff like word-processing personnel on short-term contracts) or suppliers of photocopiers, typewriters, computers and fax machines. Proximity permits intense contacts between these sub-sectors and promotes links with trade-union and other professional bodies.

These four sub-systems are very highly interdependent: single production chains comprise a sequence of administration, liaison with banks or insurers, _____ (R & D) material imports, wholesaling, processing, _____ ; and maintenance. Each sub-system _____ metropolitan environment, the UK _____ sed later. Here, office functions and

London as a Command and Control Centre

London is a pre-eminent control centre localizing, in 1989, headquarters of most of the 190 UK-owned firms in the top 500 European firms. Their combined market capitalization is US$531 billion (*Financial Times*, 19 December 1989), or US$602 billion if the two Anglo-Dutch firms – Royal Dutch/Shell and Unilever – are added, placing it far ahead of any city in the German Federal Republic or in France, where firms in Europe's top 500 have respective market capital of US$178 and US$135 billion. London also outshines New York where, since 1970, the number of headquarters of *Fortune* 500 industrial firms fell sharply from 118 to 48. Even if the *Fortune* 500 service companies are included, only 110 of the 1000 largest US firms are headquartered in New York. That does not compare with London which, in 1988/9, had 41% of the headquarters of *The Times* 1000 firms in the UK, though this represented a significant decline from 53% in 1971/2 (Westaway, 1974). South East England concentrates no fewer than 599 firms, virtually 60%. Of these, 54% engage in manufacturing as their core business and 46% in services, a picture which contrasts with employment across the region.

Table 4.1 shows that the headquarters of big manufacturing firms outnumber those of large service firms in all zones of the region, except central London, where service firms have 53% of large firm offices. Big manufacturers have 57% of all headquarters in the rest of London (i.e. inner and suburban Greater London), 61% in the outer South East, and 65% in the outer metropolitan zone.

A rich diversity of sectors is controlled by firms with London offices. Leaving aside financial services, which are pre-eminent (Chapter 3 and Chapter 5), Central London dominates commodity trading, retailing, property, advertising, hotels and leisure. Suburban and outer areas concentrate retail, civil engineering, and vehicle trading headquarters, often by extensive storage sites. All manufacturing sectors are present. Most common in Central London are offices of companies handling energy, especially oil, like BP, Shell, Exxon or Mobil, as well as British Gas, British Coal and the Electricity Council. Food, drink and tobacco firms are prominent, with Allied Lyons, BAT, Beatrice Swift, Cadbury Schweppes, Rothmans and Seagram. Others headquartered here include Imperial Chemicals Industries, Albright and Wilson, and Laporte in chemicals, Courtaulds in man-made fibres, and firms in printing and publishing (News International, Associated Newspapers), engineering (Vickers), metal (Rio Tinto Zinc, Inco Europe), electrical and electronics (General Electric Company, Standard Telephones and Cables, Thorn-EMI or Philips NV), and large holding companies like Hanson or Lonrho.

Suburban London is a stronghold of US food-processing firms like Heinz, Coca-Cola, Quaker Oats or Gerber; electrical/electronics (Hoover, MK Electrics, Plessey); and pharmaceuticals (Beecham, Glaxo, Merck-Sharpe and Dohme, Squibb). The outer zone hosts offices of engineering, chemical, metal goods, electronics (Electrolux, Hewlett-Packard, Racal) and automotive firms (Ford, General Motors, Borg-Warner and Dana).

The foregoing suggests, and Table 4.2 confirms, a major presence of foreign multinational enterprises among the largest firms (by sales volume) in the UK in or near London. In 1988/9 a third of *The Times* 1000 firms were foreign-owned, with 20% under US control. Services (excluding finance) exhibit less foreign penetration (22% of large firms) than manufacturing (42%). The latter figure partly suggests a competitive weakness of UK and London manufacturers in global markets, but the former hints less of the competitiveness of UK-owned service firms, than of monopolistic protection afforded by distance or insulation from world competition by the need for close contact with buyers (the dominance of foreign companies in vehicle distribution is an exception, but this, too, reflects weak UK-based producers and a market strength of foreign firms, like Renault, Citroen, Saab or BMW). Other European firms are active in commodity trading as American firms are in advertising, both of which stress London's international role.

The Manufacturing Sub-system

A new geography of London manufacturing is sorely needed to unravel whether recent change suggests departures from past trends, reinterpretations of well-known features or neglected facets of the capital's activities. Pioneering on these fronts demands painstaking research into the operations of firms far beyond the scope of this chapter.

A pessimistic stance would be that London's current manufacturing story can be written on a postage stamp for display in the London Postal Museum. To an extent, Peter Hall's book, *London 2001*, lends credence to this view, devoting

Table 4.1 Distribution of headquarters of the largest 1000 'industrial' firms in the UK located in London and the South East by sector and zone, 1988/9

Activity and sector	Zones						
	CL	RL	London	OM	OSE	Rest of SE	Total SE
Manufacturing – total	*146*	*58*	*204*	*85*	*35*	*120*	*324*
Oil, gas, coal and electricity	20	–	20	2	–	2	22
Metals refining and processing	12	1	13	5	–	5	18
Non-metallic minerals, building materials	2	3	5	4	2	6	11
Chemicals, fibres, rubber, tyres, plastics	16	7	23	11	2	13	36
Pharmaceuticals	6	4	10	5	2	7	17
Engineering and metal products	12	7	19	15	5	20	39
Office machinery and information systems	6	5	11	10	3	13	24
Electrical and electronics	10	11	21	12	6	18	39
Motor vehicles, components, other transport	7	3	10	7	–	7	17
Instruments and tools	2	1	3	2	2	4	7
Food, drinks and tobacco	19	12	31	6	4	10	41
Textiles, clothing, leather and shoe	5	2	7	1	–	1	8
Paper, paper products, furniture	4	1	5	3	3	6	11
Printing, publishing	15	1	16	1	4	5	21
Industrial holding, miscellaneous	10	–	10	2	1	3	13
Services – total (excluding financial and legal services)	*164*	*44*	*208*	*45*	*22*	*67*	*275*
Commodity traders and brokers	49	8	57	8	4	12	69
Retailing and wholesaling	19	12	31	3	3	6	37
Civil engineering and builders supplies	10	11	21	8	3	11	32
Property and development	20	3	23	–	–	–	23
Management, consulting, accounting	4	–	4	–	–	–	4
Advertising	18	2	20	–	–	–	20
Research and industrial services	8	2	10	6	4	10	20
Hotels and leisure activities	18	3	21	2	1	3	24

Transport and communications	14	–	14	5	1	6	20
Vehicle traders, distributors, rentals	4	3	7	13	6	19	26
Total	310	102	412	130	57	187	599

Source: The Times 1000, 1988/9

CL   Central London (defined as the City of London and postal districts WC1, WC2, W1, W2, SW1, SW2 and SE1)

RL   Rest of Greater London

OMA   Outer metropolitan area (defined as Hertfordshire and Surrey, plus the following districts – Luton, South Bedfordshire, Bracknell, Reading, Slough, Windsor and Maidenhead, Wokingham, Beaconsfield, Chiltern, Wycombe, Basildon, Brentwood, Castle Point, Chelmsford, Epping Forest, Harlow, Rochford, Southend, Thurrock, Hart, Rushmoor, Dartford, Gillingham, Gravesham, Maidstone, Medway, Sevenoaks, Tonbridge and Malling, Tunbridge Wells, Crawley, Horsham and Mid-Sussex)

OSE   Outer South East (defined as East Sussex, Isle of Wight, Oxfordshire and the following districts – Bedford, Mid-Bedfordshire, Newbury, Aylesbury Vale, Milton Keynes, Braintree, Colchester, Maldon, Tendring, Uttlesford, Basingstoke, East Hampshire, Eastleigh, Fareham, Gosport, Havant, New Forest, Portsmouth, Southampton, Test Valley, Winchester, Ashford, Canterbury, Dover, Shepway, Swale, Thanet, Adur, Arun, Chichester and Worthing)

Table 4.2 Importance of foreign-owned multinational enterprises in the top 1000 'industrial' companies in the UK located in London and the South East by sector and origin, 1988/89

Activity and sector	Greater London				Rest of South East				% foreign composition			Total Foreign
	USA	Europe	Pac△	Total	USA	Europe	Pac△	Total	USA	Europe	Pac△	
Manufacturing – total	*48*	*18*	*7*	*73*	*46*	*13*	*4*	*63*	*69*	*23*	*8*	*136*
Oil, gas, coal and electricity	6	2	–	8	–	1	–	1	67	33	–	9
Metal refining etc.	2	1	–	3	1	1	1	3	50	25	25	6
Non-metallic minerals	1	1	1	3	1	–	–	1	50	25	25	4
Chemicals etc.	7	6	–	13	6	3	–	9	59	41	–	22
Pharmaceuticals	4	2	–	6	6	–	–	6	83	17	–	12
Engineering etc.	4	–	–	4	3	3	–	6	70	30	–	10
Office machinery etc.	5	2	1	8	6	–	–	6	79	14	7	14
Electrical/electronics	6	2	4	12	8	3	2	13	56	20	24	25
Motor vehicles etc.	1	–	–	1	5	–	1	6	86	–	14	7
Instruments etc.	–	–	–	–	2	–	–	2	100	–	–	2
Food, drink, tobacco	5	2	–	7	6	–	–	6	85	15	–	13
Textiles, clothing, etc.	1	–	–	1	–	–	–	–	100	–	–	1
Paper, furniture etc.	–	–	–	–	3	1	–	4	75	25	–	4
Printing, publishing etc.	2	–	2	4	–	–	–	–	50	–	50	4
Industrial holding etc.	2	–	–	2	–	–	–	–	100	–	–	2
Services – total†	*19*	*16*	*7*	*44*	*5*	*11*	*1*	*17*	*39*	*44*	*17*	*61*
Commodity brokers	10	7	4	23*	–	1	–	1	42	29	17	24
Retailing etc.	2	2	–	4	1	1	–	2	50	50	–	6
Advertising	4	–	–	4	–	–	–	–	100	–	–	4
Research etc.	1	2	–	3	3	–	–	3	67	33	–	6
Hotels, leisure etc.	1	1	1	3	–	1	–	1	25	50	25	4
Transport/communications	–	–	1	1	–	1	1	2	–	33	67	3
Vehicle trading etc.	1	4	1	6	1	7	–	8	14	79	7	14
Total	67	34	14	117	51	24	5	80	NA	NA	NA	197

Source: as for Table 4.1

Note: * Includes also a Middle Eastern and a South African company.
          † There are no foreign-owned firms in the top 1000 in civil engineering, property or management consultancy.
          △ Pac. refers to the Asian Pacific Rim plus Australasia.

few remarks to manufacturing under 'the deindustrialization of London' (1989, 51). If that were the whole story, this chapter would not be written except to reinterpret historic glories or recent tragedies. Much is still lacking in our knowledge of patterns and processes of deindustrialization and their links with the growth of the service and information economies. At this juncture, I am reminded of a lecture I presented in 1983 at Erasmus University, Rotterdam, which elaborated some complexities of deindustrialization emerging in northwest Europe. In the discussion, an economist stated that he saw nothing complicated in analyzing deindustrialization: one could simply change the 'plus signs' in growth, multiplier or input-output models to 'negative signs' to give a correct answer! Clearly he had no notion of a real world of firms of different sizes, structures, behaviour, performance or linkages, or of the symbiosis that might coexist within a region, a city, an industrial estate, or even a single street.

What is really needed is an in-depth study of industrial change in London during the past 25 years to extend and update Peter Hall's (1962) *Industries of London* and John Martin's (1966) *Greater London: An Industrial Geography*. Such a study still does not exist because of the size, diversity and complexity of change in city industrial areas since 1970. It needs teams of researchers to document births, deaths, relocations, *in situ* changes in jobs, occupation, volume and type of output, function and linkages in the metropolis and beyond. But who would pay for such research? The 1980s witnessed sharp cutbacks on research funding and government monitoring in Britain. Small wonder that it is almost impossible to obtain up-to-date and reliable data; in many respects, the official data discussed in this chapter are testimony to the poverty and paucity of UK and London data. Abolition of the GLC in 1986 has not made it easier to find comprehensive data for the metropolis: the resources, ability and commitment of London boroughs for collecting and publishing detailed data on manufacturing and related services is very variable.

International importance

Measurement of London's importance in world manufacturing can only be very rough. Data limitations are enormous because of a lack of figures on UK regional manufacturing value-added (MVA) until recently, and of shortcomings in the international comparability of data. Yet it is quite clear that London has long been, and is still, losing its former manufacturing importance in output and employment. This is so, even though 40 years on one must still agree with Martin (1966, 60), who used 1951 UK Censuses of Population and Production data, to conclude that: 'Greater London is still the first manufacturing region of the nation by any yardstick'. Since then, it has localized a shrinking share of output and jobs which, in turn, suffered major relative global decline. Hilgerdt (1945) shows that the UK produced 32% of world industrial output in 1870 but only 14% in 1914 and 9.4% in the 1930s. The United Nations Industrial Development Organization (1983) estimated that the UK created 10.6% of global MVA in 1948 but this decreased to 8.1% in 1963 and to 6% in 1975 (although this was a recession year). Latest estimates show that Britain generated no more than 4.5% of global MVA in 1985 (The World Bank, 1988).

Until the introduction of Value Added Tax (VAT), one could only guess what the contribution of regional MVAs might be to the nation's declining share of world production. Raw data on gross regional MVA, which are now published in *Regional Trends* (Central Statistical Office, 1989), show that Greater London produced £8.37 billion MVA in 1986, or 10.3% of the UK gross MVA (£81.0

Table 4.3 London's international manufacturing importance – some comparisons of 1975 and 1986

	1975 GMO†		1975 Employment		1986 GMO		1986 Employment	
	US$bill	% of EC	000s	% of EC	US$bill	% of EC	000s	% of EC
Greater London	8.8	2.0	836	2.2	10.8	1.7	568	1.8
South East	18.0	4.1	1,741	4.5	29.7	4.6	1,487	4.8
United Kingdom	64.5	14.7	8,136	21.0	104.9	16.3	5,422	17.4
European Community*	438.1	100.9	35,657	100.0	644.0	100.0	31,119	100.0
Hong Kong	2.0		765		9.2		933	
Singapore	1.4		143		4.6		293	

Source: The World Bank (1988)
The Economist, *World in Figures, 1988*, London.
Central Statistical Office (1989)

Note: †GMO Gross Manufacturing Output.
*Calculations were made for the current 12 member states of the EC but note that the data for Greece, Ireland, Portugal and Spain included mining and utilities as well as manufacturing.

billion), with figures for the South East standing at £22.9 billion or 28.3% (UK Central Statistical Office, 1989, 55, 145). So, London still produces 0.5% of world MVA, though that compares with 8% in 1870.

The author attempted to establish the city's rank as a manufacturing centre within the European Community. The EC publishes no useful data by which to assess MVA by regions, so other methods were used. Data published by The World Bank (1988) on manufacturing product at current local prices were computed into US dollars for two years (1975, 1986) for which there are London data. National data were corroborated by The Economist's *World in Figures* (1987). The results, in Table 4.3, show that London's share of EC gross manufacturing output (GMO) slid from 2% to 1.7% in the decade, while the South East rose from 4.1% to 4.5%, and the UK from 14.7% to 16.3%. These show London's relative deindustrialization and the diffusion or diversion of manufacturing to the rest of the South East and UK. But problems arise in comparing trends in the UK with other EC nations: UK improvement is exaggerated because 1975 was a recessionary trough, while 1986 lay near the peak of a boom. Also, aggregate EC data are distorted by contrary currency-exchange movements in member countries against the US dollar and by random factors affecting a specific exchange rate in a specific year. So Table 4.3 widens the assessment to include two other world cities for which useful data are available: namely, the 'city states' of Hong Kong and Singapore. As 'newly-industrializing' cities, they exhibit the opposite trends to London, having raised their GMO far faster than the EC or South East England: Singapore by more than twice, Hong Kong by more than three times as fast. Manufacturing jobs also increased, unlike in the EC where they declined after 1975 (except in Greece and Portugal).

Deindustrialization?

London's apparent deindustrialization is a symptom of the city's (and the UK's) loss of manufacturing competitiveness on a global scale. Table 4.4 shows changes in numbers of manufacturing establishments and their employment in the two periods of 1975-82 and 1985-88. Whatever difficulties exist with these data (see Table 4.4), two trends are clear: there has been a decline in numbers of units and jobs in Greater London; and the role of the outer metropolitan and outer South East zones has strengthened. Between 1975 and 1982, Greater London suffered an unprecedented 60% decline in factories employing less than 20 people, and a 40% drop in the jobs they provided. This suggests that many plants closed, and rationalization or concentration occurred on specific sites, new and old, raising average London plant size from 82 to 123 workers (Table 4.5). Decline continued more slowly after 1985, with jobs contracting faster than the numbers of units. Data deficiencies mask how many medium-sized or small firms reduced employment to below 20 but survived. Hence the 1982-85 'jump' in numbers of units could create an illusion that government policies to stimulate small business growth and entrepreneurship had positive effects in creating more industrial jobs. That 1985 data reveal five-to-six times more units employing fewer than 20 people than those with more than 20 workers implies that the 1975-82 data hide a huge 'underbelly' of small London firms whose histories are unknown. The new data set of the 'boom' years shows a reduction in average plant employment from 21 to 19.

Table 4.6 confirms that London plants employ a shrinking proportion of UK industrial workers, but other facts are more revealing. First, its relative decline was less during the 1975-82 'deindustrialization' than in the 1985-88 'boom': so

Table 4.4   Changes in manufacturing units and employment in London, the South East and the UK for 1975–82 and 1985–88

Area	1975 A	1975 B	1982 A	1982 B	1975–82 % change A	1975–82 % change B	1985 A	1985 B	1988 A	1988 B	1985–88 % change A	1985–88 % change B
Greater London	8.83	721.73	3.55	436.51	−59.8	−39.5	23.93	506.67	23.69	455.18	−1.00	−10.16
Outer metropolitan	5.32	572.90	2.89	417.10	−45.7	−27.2	16.08	469.80	16.71	449.86	+3.96	−4.2
Outer South East	3.56	383.53	2.15	286.25	−39.7	−25.4	12.23	322.64	13.04	337.81	+6.60	+4.7
Total South East	17.71	1,678.17	8.58	1,139.85	−51.6	−32.1	52.24	1,299.12	53.44	1,242.84	+2.30	−4.33
United Kingdom	117.85	7,157.59	33.01	4,645.14	−72.0	−35.1	150.04	4,936.89	154.47	4,849.00	+2.95	−1.78

Source: *Business Monitor*, PA1003, 1976–89 (DTI, Business Statistics Office)

A   Number of legal manufacturing units (000s)
B   Number of employees (000s)

Note: Data for the two periods 1975–82 and 1985–88 are not comparable. Figures published for years up to and including 1982 exclude all manufacturing units engaging 1–19 employees because 'information relating to these smaller units is of doubtful reliability' (*Business Monitor*, PA1003, 1982, p.3) while data for 1985 and subsequent years include 'sites with employment below 20' belonging to firms registered for VAT purposes. No data was published for 1983 and 1984 while the new data bases were being prepared.

Table 4.6   Manufacturing units and employment in London and the South East as a percentage of the UK total, 1975–82 and 1983–88

Area	1975 Units	Employment	1982 Units	Employment	1983 Units	Employment	1988 Units	Employment
Greater London	7.5	10.1	10.7	9.4	15.9	10.3	15.3	9.4
Outer metropolitan	4.5	8.0	8.8	9.0	10.7	9.5	10.8	9.3
Outer South East	3.0	5.4	6.5	6.2	8.2	6.5	8.5	7.0
Total South East	15.0	23.5	26.0	24.6	34.8	26.3	34.6	25.7

Source: *Business Monitor*, PA1003, 1976–89

Note: All percentages are rounded so totals do not necessarily add up to 100

Table 4.5 Changes in the number of employees per establishment in London and the South East, 1975–82 and 1985–88

Area	1975	1982	1985	1988
Greater London	82	123	21	19
Outer metropolitan	108	144	29	27
Outer South East	108	133	26	26
United Kingdom	61	141	33	31

Source: *Business Monitor*, PA1003, 1976–89

Note: All figures are rounded

London did not suffer faster deindustrialization than the whole UK. Second, the metropolis actually raised its share of UK medium and larger-sized manufacturing units from 7.5% in 1975 to 10.7% in 1982, implying that such firms in London were more competitive and better able to survive *national* deindustrialization. Third, between 1985 and 1988, the proportion of VAT- registered manufacturing units in Greater London declined, suggesting that a higher closure rate than in the UK explains the faster decline in the city's share of UK manufacturing employment in the boom years. Clearly, more than business cycles were affecting London manufacturing trends. Fourth, the South East increased its concentration of UK manufacturing units from 15.3% to 26% between 1975 and 1982 largely because medium-sized plants employing more than 20 workers were established or survived. However, as their share of UK jobs remained stable, this indicates that they employed fewer workers in the outer metropolitan and South East subregions in these years. Recently (1985-88), numbers of units and jobs have stabilized in the outer metropolitan area, but the outer South East is still increasing its share of UK plants and jobs.

Overall, the mid-1980s 'boom' witnessed accelerated industrial contraction in plants and jobs in Greater London relative to the UK. This process is now diffusing into the outer metropolitan area. Continued, if slower, growth in the outer South East explains why the metropolitan region still retains one quarter of the UK's manufacturing sector employees in 1988. Thus, broadly, manufacturing in the South East increased its 'survivability', if not also its competitiveness, in the UK and international economy in the 1975-82 period and retained it through 1985-88. Yet 'survivability' was least in Greater London and increased with distance from the metropolis, though it weakened generally and diffused further into non-metropolitan areas as the 1980s progressed. Reasons for this will be suggested later.

These findings are confirmed by Table 4.7: the 1975-82 decline in numbers of factories employing more than 20 workers, and of jobs, was less severe in outer South East counties (Oxford, Buckinghamshire, Berkshire, West Sussex and Hampshire) than elsewhere. Between 1985 and 1988, though, all counties, except Greater London, saw growth in numbers of units which was more rapid in the less-industrial outer South East, although only Sussex and Hampshire recorded job growth. Manufacturing jobs shrank everywhere in the outer metropolitan zone, especially in the north and west, including the former 'boom' county of Berkshire.

Table 4.5 sheds more light on these trends, though the unfortunate absence of output, value-added and investment data at regional level means we must use employment as a proxy for size, an indicator which has often caused serious

Table 4.7 Manufacturing changes in London and counties in the South East, 1975–82 and 1985–88

Area	1975 A	1975 B	1982 A	1982 B	1975–82 % change A	1975–82 % change B	1985 A	1985 B	1988 A	1988 B	1985–88 % change A	1985–88 % change B
Greater London	8.8	721.7	3.5	436.5	-59.8	-39.5	23.9	506.7	23.7	455.2	-1.0	-10.2
Essex	1.2	129.6	0.7	93.8	-43.1	-27.6	4.1	109.3	4.4	101.5	+8.9	-7.1
Bedfordshire	0.6	76.4	0.3	62.0	-49.3	-18.8	1.6	64.0	1.7	58.2	+2.5	-9.1
Hertfordshire	1.2	134.8	0.6	89.6	-48.2	-33.5	3.2	100.7	3.3	100.3	+2.9	-3.8
Buckinghamshire	0.7	59.4	0.4	47.8	-40.4	-19.5	2.2	61.5	2.4	60.3	+8.6	-2.0
Oxfordshire	0.3	49.7	0.2	35.4	-30.5	-28.8	1.3	29.5	1.4	40.4	+10.2	+37.1
Berkshire	0.7	76.9	0.4	59.5	-43.9	-22.7	2.2	64.2	2.4	57.6	+7.7	-10.3
Surrey	0.9	67.8	0.5	47.5	-46.5	-29.9	2.8	60.0	2.9	54.9	+4.6	-8.6
Hampshire	1.1	136.9	0.7	107.5	-41.3	-21.5	3.7	119.9	4.0	121.5	+10.0	+1.3
W. Sussex	0.6	55.2	0.4	45.5	-41.1	-17.6	1.9	50.8	2.0	52.6	+4.6	+3.5
E. Sussex	0.4	36.9	0.2	23.7	-45.4	-35.6	1.6	28.6	1.8	29.0	+9.3	+1.5
Kent	1.1	123.5	0.7	84.3	-42.6	-31.7	3.4	96.6	3.7	93.3	+8.1	-3.4

Source: *Business Monitor*, PA1003, 1976–89 (DTI, Business Statistics Office)

A Number of legal manufacturing units (000s)
B Number of employees (000s)

Note: Data for the two periods 1975–82 and 1985–88 are not comparable. Figures published for years up to and including 1982 exclude all manufacturing units engaging 1–19 employees because 'information relating to these smaller units is of doubtful reliability' (*Business Monitor*, PA1003, 1982, p.3) while data for 1985 and subsequent years include 'sites with employment below 20' belonging to firms registered for VAT purposes. No data was published for 1983 and 1984 while the new data bases were being prepared.

misconceptions about the role of firms of different sizes in economic decline or revival. Obviously, plants surviving the years 1975-82 were significantly larger employers throughout the UK and the South East (i.e. amongst plants employing more than 20 workers). That Greater London and the South East raised their share of UK survivors in these years may be linked to their above-average size in all three sub-zones, implying that manufacturing operated at larger, more efficient scales than elsewhere in the UK. But the South East suffered less recession in these years: a more favourable regional market environment could have helped local units to employ more workers, especially if they were supplying the expanding financial and office markets of London and the outer metropolitan zone or more efficient, surviving firms elsewhere in the UK or continental Europe. Research is needed to investigate these hypotheses.

The more recent data, however, tell quite a different story. Average unit employment size declined, which could reflect job losses by medium and larger units, birth of new units, or the greater importance of firms employing less than 20 workers. Units in the South East are now 'below average' in job terms relative to the UK, which may be due as much to greater efficiency, higher capital/labour and higher output/labour ratios, as to a growth of small firms. There is no published evidence that Greater London firms are prominent in high-technology sectors, in marked contrast to the outer metropolitan zone (*Financial Times*, 16 February, 1990).

Much less has been made of this view in the past 15-20 years as large factories owned by larger firms have closed or shed labour in large numbers. Greater London has many examples, yet Table 4.5 shows it has the smallest average-size for firms in both 1975-82 and 1985-88. As this is also the zone with the most significant decline in job numbers and in its share of UK manufacturing jobs, smaller size appears to be a weakness. This may express a lack of viability of small firms in the metropolis or that firms in surviving older sectors have contracted so much in number that the necessary 'critical mass' for local linkages, external economies, market share and investment capability has fallen below the survival threshold. Sectoral data are not currently to hand to test this hypothesis for these years.

London's Competitiveness in the International Economy

During the past 30 years, the city-region was transformed from a dominantly materials-based, to a mainly information and financial-based, transactional economy. These structural shifts have been, and still are, interdependent. To an extent, the deindustrialization of London is a function of the inability of the materials-based productive sub-system to compete for land, labour, services and infrastructure with administrative, information and financial-services subsystems. Stated another way, manufacturing suffers a weakening potential to compete in national and international markets from sites in Greater London, but retains higher capability to do so from places in the outer metropolitan and South East zones. In this respect, London resembles New York City which continues to lose manufacturing jobs in larger numbers than the financial sector. *Fortune* (February 1990) considers this ironic: manufacturing moves out less because New York is declining, more because bankers, brokers, lawyers and other service firms find it the best place for business and bid up average downtown office rents. Is this also true in London?

Restructuring the London regional economy results from interactions between populations of firms and environments at three levels: metropolitan-regional, national and international. Globalization has accentuated the supply-side of the economy relative to the demand-side, and thus re-emphasized the importance of international competition between, and hence the comparative advantages of, cities and regions. It is completely incorrect, however, to conceive of 'comparative advantage' in single factors such as wage rates. For any place, and especially a complex metropolitan regional economy like London's, a kaleidoscope of factors interact and intermesh to create a 'basket' of comparative advantages and disadvantages. These 'baskets' clearly differ between its sub-regions: the central city, inner city, suburbs, or outer metropolitan zone.

Each 'basket' comprises populations of firms and the regional environment, but it is helpful to disaggregate these for further analysis. Porter (1990) suggests that 'comparative advantage' has four sides. Two are embedded in the population of organizations: that is, they comprise the strategies, structures and rivalries of firms, and their related or supporting clusters of activities. The other two are integral to the environment: viz. demand and factor-supply conditions. These four will now be examined to assess London's competitiveness in the international economy.

Firms

The question that arises here is whether London's manufacturing firms exhibit weaknesses which speed metropolitan deindustrialization by making them less competitive with regard to both UK and foreign markets for manufactures, and service activities within the region? In fact, such weaknesses do exist and are related to technological backwardness and dependency. This is because London exhibits tendencies towards economic dependency as a manufacturing, services and headquarters city. That may appear to contradict its role as a core-imperial and decision-control centre in the global system. Yet dependency arises out of the penetration of UK and metropolitan markets by foreign multinationals which, in turn, reflects a limited tradition of control over technological innovation by leading London-based firms. Gin, the spirit drunk by the working classes in Hogarth's time, is highly symbolic of London industry: the city's indigenous manufacturing evolved to specialize more in low value-added, mass consumer goods. The capital has no really internationally-renowned firms to rival the highly-valued and prestigious perfumes, fashion clothes or Citroen cars of Paris, for instance. Famous names, like Yardley or Maudslay, have been absorbed into the subdivisions of large, acquiring firms like BAT, or have disappeared into oblivion.

The point is this. The lesson to be drawn from the post-war world economy is that export-base theory (North, 1955) is still relevant at every geographic scale, but one key modification is required for a metropolitan region like London: that prosperity can *only* be assured if indigenous firms yield large *net exports* of high value-added, technologically-based manufactures for sustained periods. The example set earlier by the USA and, more recently, by the German Federal Republic, Japan, South Korea and Taiwan, support this. And firms must be able to raise product and processes standards regularly to counter dynamic shifts in international comparative advantages (Linge and Hamilton, 1981). The argument that producer or consumer services can replace manufacturing to create export surpluses is open to much doubt because such services need proximity to consumers and hence their international tradeability is far lower than for manufactures. London's problem can be illuminated by examining the contrasting roles of British and foreign manufacturing firms.

British firms

A substantial market or niche specialization exists between UK-owned and foreign-owned multinationals. Stopford and Dunning (1983) first diagnosed this on a global scale. Many key UK firms which are headquartered in London, including those with deep manufacturing roots there, are renowned for their firm-specific ownership of marketing advantages – which have been nurtured in the city's imperial commercial traditions – rather than of superior technology. Not surprisingly, such multinationals are prominent in the more material- or labour-intensive sectors which endowed the UK with competitive advantages in the first and second industrial revolutions and were consolidated through trade management in 19th-century imperial expansion.

Firms in most of these sectors have run down manufacturing in London or limited their activities there to headquarters' administration of national and global business. The reasons are not hard to seek. First, most such firms make mature, lower value-added products, which draw on lower labour skills and are highly vulnerable to competitive displacement from world markets by low-cost producers in the Third World or NICs. Second, as long as firms retain manufacturing in the UK, the high degree of technical standardization in their products means that they have to relocate processing facilities to regions with low labour costs. Third, the materials-handling and marketing ends of their businesses generally require a strong presence overseas where most turnover is created: for Unilever, for instance, the UK is one of 180 markets. Fourth, their products do not usually require advanced Research and Development in the London region; such needs are often best met at UK manufacturing sites in other areas or abroad in material-rich or environmentally-different regions. And fifth, some big UK multinationals, like BTR or Lonrho, became conglomerates or holding companies, usually for three reasons: to reduce their exposure *as firms* to competition by diversifying and spreading risks across a range of products, services and across geographic markets in both developed and developing countries; to exploit their firm-specific trading and managerial advantages by differentiating products; and to derive added strength from London's pre-eminence in advertising, information and financial services.

Thus one reason for London's manufacturing decline is a general absence of technologically-strong local firms capable of reinvigorating the economy in the 1980s. Exceptions are pharmaceuticals, with firms like Beecham, Glaxo and Wellcome, and, to a lesser extent, electrical-electronics firms of Thorn-EMI and ICL. ICI is another example, but its main base is elsewhere. Symptomatic of London's problem is ICL. At the time of writing ICL is the only UK-owned computer firm ranking in the top seven manufacturers in Europe; but it lies 5th, achieving a sales volume in 1989 of just 39% of Siemen's computer division, and 2% of IBM's (*Financial Times*, 26 February 1990).

Foreign multinationals

Foreign firms dominate newer sectors of the third and fourth industrial revolutions which involve greater technological sophistication and closer integration of management, R & D, manufacturing, assembly and marketing. These firms became the key *force motrice* in London's 20th century industrial growth and diversification. The metropolis became quite dependent on foreign firms for employment, wealth-creation and economic stability in a widening range of sectors such as oil-refining, petrochemicals, chemicals, electrical-electronics products, transport equipment (especially motor vehicles), rubber, office and data processing machinery, machine tools and pharmaceuticals (see Table 4.2).

Table 4.8 Aggregate changes in the number of UK firms and foreign-owned multinational enterprises in manufacturing headquarters in London and the South East among the top 1000 'industrial' firms in the UK 1978/9–1988/9

	1978/9 Number		1988/9 Number	
	U.K.	Foreign	U.K.	Foreign
Central London	119	44	102	44
Rest of Greater London	33	22	29	29
Sub-total: London	152	66	131	73
Outer metropolitan	25	27	38	47
Outer South East	9	5	19	16
Rest of the UK	166	24	183	25
Total UK	342	132	371	161

Source: *The Times 1000*, 1978/9, 1988/9

Foreign multinationals contributed 21% of the GMVA in the South East in 1986 (and 17% in the UK). This figure masked their real concentration in the London region where they yielded 35% of all GMVA by foreign firms in Britain, although both these figures were below those in 1981 (which stood at 25% and 39% respectively). That said, foreign manufacturing firms became more prominent among larger firms headquartered in all zones of the city region between 1978/9 and 1988/9 (Table 4.8). Indeed, the proportion of foreign multinational corporations among large manufacturing firms located in the South East rose from 34.5% to 39.5% in these years. Additionally, the number of the big UK manufacturers headquartered in Central London shrank more sharply than those of foreign firms with offices there, while in the other zones, as UK firms declined, foreign companies rose in number. Today, big foreign multinationals outnumber large UK-owned firms in the outer metropolitan zone where they have become more localized relative to the rest of the UK. Thus foreign firms compete more effectively for local resources and can pay higher rents and wages as they are more efficient, use more modern technology and have the financial backing of stronger currencies than sterling.

The long history of foreign investment in London manufacturing (Chapter 2) stresses its early dependency on imported technology. Siemens opened a factory on Millbank in 1858 to make instruments and batteries for the British government. Hoe (New York) set up the first US-owned branch in the UK in 1867 to manufacture its new revolving printing presses in London for Lloyds and *The Times*; soon they were sold to all newspapers in the British Isles. And French technology shaped the early motor industry, with Darracq assembling cars in Fulham from 1911, and Citroen in Feltham in 1922.

But it has been the supremacy of new US product and process technologies, evolved in the world's biggest home market that was undamaged by war, that has left the deepest imprint. Dunning (1958) identified four stages in inward US investment to Britain. These are readily discernible in London: the formative years to 1914; steady expansion between 1919 and 1929; a major upswing from 1929 to 1939; and a growing importance from 1940 to 1960. US firms penetrated UK markets with a widening array of goods made in plants located in or near the metropolis. Of 306 US firms in the UK in 1953, 42% (129) were in London or within a 50km radius of its docks, for easy access to imported materials, parts or sub-assemblies from the USA and to export to non-dollar world markets (Dunning, 1958). They employed 115 700 people, 47% of all those in US firms in Britain, and 25% of all industrial workers in London and the South East, a proportion far higher than elsewhere (the North West at 14% and Scotland at 10% were the next most dependent). Dunning (1958, 86) notes that '... between 1925 and 1938, nearly forty American branch manufacturing units were set up ... on or around the ... Great West Road and North Circular Road ... Welwyn Garden City and ... Slough on planned estates ...'.

Many US firms became household names like Westinghouse, Ford, Borg-Warner, GM, Firestone, Hoover, Frigidaire, Burroughs, IBM, National Cash Register, RCA, Gillette, Heinz, Mars, Kodak, Johnson & Johnson, Elizabeth Arden or Revlon. In all cases, the large city market and good transport access to other UK regions were decisive in attracting them to London. Yet many chose to use the city also as a production platform for manufactured exports to Europe, and to British colonies (later the Commonwealth) under the label 'Made in England' in order to gain access to preferential UK export markets which they could not supply from the USA.

Symptomatic of UK market penetration by foreign firms was the growth of pharmaceuticals output after 1945 to supply the National Health Service. Dunning (1958) found that 24 manufacturers in the UK were selling ethical drugs to the NHS: only seven were UK-owned while 13 were American and three Swiss. Most operated in or near London because of high quality R & D facilities locally. Examples were Merck-Sharpe and Dohme, Squibb, Eli Lilly, Abbott, Searle and Ciba-Geigy, while others like Sterling Health, Bristol-Myers and Vick produced a wide range of over-the-counter medicines.

By the 1970s, the US-owned plants were ageing and in need of replacement. The test became whether London could retain or would lose them. New factors shaped the outcome. First, the firms had become more multi-functional, multi-site organizations which could alter activities at particular locations after making comparisons with other places where they already operated or where they might open new facilities – witness Ford's comparisons in 1989 of productivity at Dagenham with that at its plants in Valencia (Spain) or Genk (Belgium). Second, capital was far more mobile and permitted easier transfer of functions and capacities from places with high and rising costs or dull markets to places with low costs or growing markets. Third, Britain's entry into the EC changed some US managers' perceptions of London as a base from which to supply that enlarging market. Fourth, sharper competition followed the first oil crisis as the world economy evolved from a uni-centric pattern of dominance by large US firms to a tri-centric one of large multinationals based in North America, Western Europe and Japan. Rising import penetration of the UK market by European and Japanese firms put pressure on US firms to replace old and adopt new processes and products to cut labour costs and reorganize their operations on a European-wide scale. Fifth, as recession deepened, national governments and the EC offered more financial aid to steer new investment away from metropolitan areas to peripheral and industrially-declining regions. Finally, new technologies – microprocessors, robots, computer-assisted design and manufacturing, and global telecommunications – made radical change possible.

These forces initiated far-reaching restructuring by foreign multinationals in London between 1975 and 1990. Two broad changes can be identified. First, many existing firms closed, slimmed down, or changed functions in, their older London units. Various scenarios emerged in this deindustrialization. Some, like

Monsanto, divested their European plants but retained sales offices in London. Others, like Hoover, Xerox or Motorola, relocated to, or expanded output in, new plants in assisted areas like Scotland or Wales to tap government grants or pay lower wages. Xerox and Hoover also shifted output to continental Europe to be more centrally located in the rapidly growing EC market. Others, like IBM (Bakis, 1987) or Electrolux, rationalized on a pan-European basis. Ford has combined options and effectively 'disintegrated' the Dagenham works. The company turned London from a net exporter, into a net importer, of vehicles and parts. A generation ago, Dagenham made steel and an entire range of cars (Escort, Cortina, Capri, Granada) for the UK and export. Step by step, Ford transferred parts output to new plants in Bridgend, Strasbourg or Bordeaux, and the assembly of middle- and up-market cars (Sierra, Granada) to Belgium and West Germany, so all these models must now be imported. The London operations have been polarized into assembly of small, low-value Fiestas and Escorts at Dagenham, and R & D and management of European business near Brentwood (in the outer metropolitan zone). In many respects, this reflects the polarized metropolitan regional environment to be discussed later.

Second, from the late 1970s the inward movement has been accelerated by a new generation of US, European and Japanese firms which command sectors like information technology, microelectronics, biotechnology and pharmaceuticals. Rarely have they located in Greater London itself: most commonly, they opened new facilities integrating their European or UK offices, manufacturing and R & D in or near the so-called 'Golden Triangle' tributary to Heathrow and Gatwick airports in the west. Newcomers there in electronics include US firms such as Apple, Control Data, Digital Equipment Corporation, Sun Microsystems, Tandem, Tektronix and Wang, as well as Fujitsu and Toshiba of Japan. West Germany and Switzerland are the main sources of investment in pharmaceuticals, including firms like Bayer, Roussel Laboratories, Boehringer Ingelheim and Sandoz.

Mergers

Western Europe and North America have experienced intensified acquisition activity since 1985. Though the financial transactions were often made in New York, mergers in food-processing, pharmaceuticals, publishing and the media, are particularly relevant to London. One case illustrates this. The largest merger in 1989 was the leveraged-buy-out[2] of the US foods corporation RJR Nabisco, by Kohlberg Kravis Roberts for US$25 billion. To profit from it KKR sold off parts of RJR Nabisco's European operations. This sealed the fate of one of London's largest biscuit factories, Peek Frean's at Bermondsey, which was bought, then closed (with large job losses) by the French food firm, BSN, as it rationalized its operations into a series of new large-scale factories, mainly in France, to serve the entire European Community by 1992.

The full implications of these new trends are not yet fully understood. Yet the rate of cross-border acquisitions is rising in Europe as firms prepare for 1992. Such mergers rose in value from US$5.1 billion in the first quarter of 1989 to US$16 billion in the third. Although UK firms have long been active in this process, a significant change occurred in the international pattern in 1989 in that UK firms, or UK subsidiaries of foreign firms, became the target for European predators: half the acquisitions in the EC involved continental (and Irish) purchases of UK firms. Clearly, this threatens to weaken London's role both as a manufacturing centre and premier headquarters city in Europe.

Clusters of activities

London's national and international importance derives not simply from the large numbers of UK and foreign firms and their functions in the metropolitan region. It also depends crucially on the competitive strengths or weaknesses embodied in the region's clusters of related activities. Obviously, the four productive sub-systems outlined earlier (materials, transaction, information and administrative sub-systems) are central in this. Specifically, London's competitiveness depends on the forging of efficient linkages between suppliers and customers and on stimulating innovation among producers and consumers at various stages of each production chain, both within and between the four sub-systems.

Like the theory of comparative advantage, the concept of activity clusters is enjoying a revival. This is no accident. The rise of 'high tech' industries highlights the developmental significance of Joseph Schumpeter's 'clusters of sectors' in diffusing self-reinforcing growth in each wave of innovation. The geographic concentration of high tech production in areas like the Santa Clara ('Silicon') Valley of central California, the Thames Valley west of London, or the 'Third Italy', are rekindling interest in Alfred Marshall's 'industrial districts'. In turn, this has revived Francois Perroux's 'growth pole' ideas as a basis for policies to 'reindustrialize' deindustrializing areas.

To an extent, 19th-century clusters of furniture, printing and precision industries exemplify early innovation waves in London. Relatively long-term competitive strength emanated from the inter-war in-migration, in follow-the-leader fashion, of rival US firms which brought linked activities, especially in the automotive sector. Yet firms like Ford, GM, AC-Delco, Borg-Warner, Champion Spark Plugs and Firestone enticed other US competitors into London which stood to gain from wider car-ownership and use: those insurers, credit firms and rent-a-car firms that were already tied to Ford and GM in the USA (Hamilton, 1976). However, an absence of strong UK competition in the sector led to complacency by these firms and their failure to upgrade technology led to their recent demise or reorganization. The 1970s and 1980s in-migrants involve a broader cluster of high-tech manufacturing and producer services like software or management which is linked to R & D by universities, government laboratories and private firms.

Global competition between the key manufacturers is integral to dynamism in these production clusters, and they have concentrated such dynamism in the west of the city region. Pharmaceuticals illustrates this. Two major UK firms, Beecham and Glaxo, operate there. As a result of successful R & D to develop anti-ulcer and AIDS drugs, Glaxo is the sector's European leader. Their presence has acted as a magnet to an influx of foreign pharmaceuticals firms into nearby areas (as noted above). A US$8.3 billion merger between Beecham and SmithKline Beckman of the USA may heighten London's role as a premier European centre in this sector. The two firms claim that the merger that created the world's third largest pharmaceuticals corporation, will raise their combined revenue and cut combined costs by consolidating costly R & D, merging manufacturing plants and sales forces: rationalization will affect 60 sites worldwide.

The concept of clusters also concentrates the mind on populations of units in a production system and can thus reveal the *interdependencies* between small, medium and large organizations that make such production systems work. Two aspects of this need comment here. First, the 1980s has witnessed a policy-orientated preoccupation with the fostering of small firms (employing less than 25 workers) to counter the socio-economic effects of big job-losses by large firms (with more than 500 employees). This was rooted in a perhaps misguided conviction that small firms are more innovative and have more growth potential than large ones. In fact, earlier discussion of Tables 4.4 to 4.6 suggested that the dominance of small firms in Greater London might be a cause of its more rapid deindustrialization. The point is that London's traditional small firm sectors that survive are labour-intensive (clothing, printing and furniture) and have suffered a resurgence of sweat shops, outwork and sub-contracted home work which shade into a burgeoning informal sector with abysmal working methods (Greater London Council, 1985b). Moreover, London's small firms exhibit a chaotic foundation of unrelated activities, a dependency on unhelpful banks for credit or on unenlightened local authority advice, and have been vulnerable to high death rates. The point is that small firms are unsuccessful modes of regeneration without the support of a critical mass of connected activity and, not infrequently, the spur and stability of a large innovating organization.

Second, central to the growth and development of all clusters is harmony between all links in the production chain, irrespective of the size of firms. As Porter (1990) observes, joint problem-solving yields faster, more efficient solutions: suppliers tend to act as conduits for information transmission from company to company, thus speeding innovation or responsiveness to market changes. On the other hand, conflicts of interest between the members of a population of firms in a sub-system can weaken clusters, leading to their ultimate collapse. The Greater London Council (1985b) highlights how deindustrialization of inner-city industrial districts specializing in foods, furniture or clothing resulted from confrontation between partners with unequal bargaining power: small manufacturers against big retailers like Asda, Safeway or Sainsbury (foods), MFI (furniture) or British Home Stores (clothes) which demand sharp cost reductions. Similarly, London's radio and television firms could not match the low-cost needs of chains like Dixons. In all cases, the retailers switched orders to cheaper suppliers in other UK regions or in other countries.

Demand conditions

London's deindustrialization also reflects a shift from relatively advantageous to quite disadvantageous market conditions. Significant changes in demand since 1970, mainly towards high-technology and high-quality manufactures, have markedly tipped the balance and have revealed serious weaknesses in the regional environment which are rooted in the distant past. One symptom is revealed by the National Economic Development Council's (NEDO) monitoring of UK trade since the 1970s. The NEDO found that UK exports are relatively low value and imports have higher value-added. Yet more worrying is the fact that 'a downward drift is now an established feature' (*Financial Times*, 16 March 1990); i.e. UK firms are increasingly specializing in lower value manufactures. The following demand conditions are to blame.

The first is a sluggish growth of national demand for producer and consumer goods. This emanates from an abysmally poor investment performance of UK firms and their low rate of capital replacement which, in turn, yielded slow productivity and thus slow personal income growth as compared with all other developed economies. Both factors have limited the growth of UK market size and hence of opportunities for UK firms to achieve scale economies. And foreign firms simply import the technology they require. The problem has been compounded by governmental preoccupation with employment creation or unemployment reduction in purely quantitative terms with little or no regard for the qualitative, wealth-creating or skill aspects of jobs. As expected, Table 4.3 shows that London, the South East and the UK account for higher shares of manufacturing employment (than output) within the EC, although the serious 'overmanning' evident in 1975 had been reduced by 1986. Clearly, the city's region's firms are more labour-intensive than those in the Community.

A slow rate of capital replacement, however, deters innovation. This links with the second, and most damaging, weakness in the demand conditions, namely a low level of commitment to high-quality products and services. This, above all else, explains the general lack of technologically-orientated London firms and the dependency on overseas firms. The origins and manifestations are many. Only few can be mentioned here. Centuries of preferential access to imperial markets made most manufacturers complacent or uninformed about competitors. But such protected access brought London firms into sustained contact primarily with Third World or frontier environments which encouraged them to perpetuate the sale of low quality, outmoded manufactures. No less destructive of London industry is the 'penny-pinching' or 'cut-price' approach adopted over decades to private, public-sector and infrastructure investment in the UK. Education, health, housing, public transport, roads, sewage, water-supply and even much manufacturing investment are all regarded as 'costs' to firms or taxpayers.

That perception inevitably fosters an environment in which people provide, accept or want least-cost, low-quality goods and services. This attitude is deeply engrained in British society and levels of power. It has contributed to national decline, with negative effects on London. A classic case is the UK civil engineering and building-materials sector. This industry has long been used to build cheap private houses, shoddy 'low-cost' public housing or offices of little architectural merit. The recent boom in London's commercial property reveals the sector's near-medieval standards of design, construction methods, and labour training (from managers to labourers). The results are two-fold. Some developers were unable to find UK firms capable of constructing the custom and architect-designed offices they required and engaged foreign companies like Kumagai Gumi of Japan or Skanska of Sweden. Furthermore, building-materials firms have been unable to meet high specifications so that the boom has sucked in huge imports from continental Europe, creating a £3 billion trade deficit in 1989 in construction materials and components (*Financial Times*, 14 March 1990).

Environmental protection presents another case: a lost opportunity. After the smogs of 1953, London pioneered air-pollution control in 'smokeless' zones. Yet over the years, adoption of lax, 'cost-orientated' views by government at all levels to all forms of pollution engendered a 'couldn't care' milieu which puts no pressure on firms to innovate in this realm. Hence, initiation and enforcement of tight environmental laws will mean further UK dependence on imported equipment or on in-migrating US, Japanese or Swedish manufacturers.

All these 'costs', however, ought to be regarded as investments in future productivity and future generations. The aim should *always* be at *world leadership* in quality standards. Many of the UK's competitors adopt such a stance. Two examples illustrate that such a perspective can yield economic benefits. Strong commitments to the NHS and a healthy farm livestock economy after 1945, supported by superb state-backed R & D, have fostered a milieu for firms like Beecham, Glaxo and Wellcome to become world leaders in pharmaceutical and veterinary products. Similarly, the high standard set by the BBC, and matched by Independent Television, has been a major force in London's pre-eminence in media production: in 1988, for instance, the UK displaced the USA as the world's leading exporter of television films.

A third demand factor which has assisted London's deindustrialization is the phenomenon of so-called 'market splitting' or market fragmentation into wealthier and poorer segments. At the local scale, polarization is evident in the clothing industry, with a fashion sector in the West End near large stores and wealthier clientele, and mass, low-cost output in the East End near large working class areas in industrial and dockland London. At the metropolitan scale, a transition from the economic growth of the 1960s to stagflation in the 1970s and recession in the early 1980s was paralleled by a transition to 'more discriminating consumption' (Harvey, 1985, 215). Formerly, mass manufacturing output had enabled a large semi-skilled and skilled industrial workforce to engage in mass consumption of UK-made goods. Deindustrialization has 'split' demand through shrinkage of a middle-income market of industrial workers and pauperization of the unemployed. And services growth brought low-paid, part-time work at one end, and high-income earners in producer services at the other. Such 'market splitting' fed a vicious circle of rising import penetration and further deindustrialization. High-income groups sought custom-made, high value-added products which local and UK firms could not provide. Low-income groups sought cheap goods, thus encouraging large retailers to import electrical goods or clothes from the Third World and NICs and undercut local producers. The process also fostered demand for Do-It-Yourself (DIY) household goods, but again, foreign firms like US-owned Black and Decker or Swedish IKEA, have the competitive edge.

Supply conditions

Historically, London's economic growth, development and diversification, and its international competitiveness, were greatly assisted by plentiful, cheap land, labour and capital and, after 1840, by a dense and efficient public transport system. By the period 1870-1914, however, access to local capital was losing its advantage: as much as 40% of all UK capital was exported then as the City neglected home investment in new manufacturing sectors, preferring projects yielding quick profits overseas. Net capital export continued virtually unabated in the 20th century and the UK lost its rank as the world's second largest capital exporter (after the USA) to Japan and West Germany only in 1987.

In recent decades, several factors have combined to tip the balance in the 'basket' of regional supply conditions from comparative advantage to worsening comparative disadvantage. Currently, the most-publicized problem is a severe shortage of skilled labour, which is inflating wage costs. It is now widely recognized that the UK has the smallest proportions of, and least-well qualified, skilled labour in the developed world. This inhibits the ability of firms in Greater London to adjust to global competition by adopting new technology and now threatens to halt the high-technology boom in the Thames Valley. Everyone is to blame for the shortages: all levels of government, firms themselves, trade unions and the lack of aspirations among the population at large. The root causes are discussed above under 'demand conditions'.

Yet since the 1950s, a series of 'booms' in office and services growth in the City of London and the West End have generated land and property price spirals which have made the metropolis *the* 'inflation leader' in the UK. Most acute was land and property inflation associated with preparations for Britain's entry into the EC, when hundreds of foreign banks opened premises in the Square Mile, and the 'big bang' of 1987, when the introduction of 24-hour trading in currencies, stocks and shares unleashed a further round of expansion by financial institutions. Increasingly, therefore, business and government have attempted to depress labour costs and infrastructure 'costs' to offset high and rising land prices throughout the UK.

It was partly for these reasons that an area of Docklands was designed as an Enterprise Zone in the early 1980s. The arguments were that such a zone, free of bureaucracy and taxes, could replicate the 'free market conditions' of Hong Kong or South Korea and hence create jobs for local, semi-skilled workers. It was seen as a tool for sustaining manufacturing in Inner London by helping to match the low wages and production costs of the Third World (Talbot, 1988). The idea was misguided on three grounds. First, it ignored the very high quality of infrastructure being developed in many NICs where such zones are located. Second, such zones tend to displace firms from other places rather than create net employment (Department of Environment/Roger Tyms & Partners). And third, by encouraging low-cost manufacturing, they run the danger of trying to attract the very firms who will quickly suffer even lower-cost competition from NICs and Third World countries. Fortunately, London's Docklands has drawn in activities like newspaper firms and media establishments which must maintain close ties with Central London. Nevertheless, Docklands development is slow and expensive: the London Docklands Development Corporation estimates that 8000 jobs were created by mid-1989 at a cost of £1.2 billion, which is about £150 000 per job.

Of longer term concern, central government, irrespective of party politics, repeatedly pursued 'stop-go' policies in the post-war period to rein inflation and support sterling. The results are always the same. Cuts in public expenditure on social and technical infrastructure raise operating costs for the entire population of firms: low labour productivity ensues from inadequate health care, housing, education or training, while transport costs are raised by traffic congestion and outdated equipment. And high interest rates inflate investment costs, depress the take-up of capital by manufacturing and services and throws them into further dependence on cheap labour for survival. A vicious circle is completed as abundant UK and foreign capital shies away from 'unprofitable' manufacturing and flows into 'more profitable' property development in London, inflating prices still further. This trend is deepened by government policies to foster higher levels of home ownership.

The diffusion of phenomenal rises in land prices and property rentals from the prime sites in Central London has had several consequences. First, it has accelerated a decline of manufacturing in Greater London in the 1980s 'boom' years, driving it into the outer zones. Second, it is also now forcing many firms to vacate expensive office space in the central area and relocate, especially to areas west of London close to the airports. Reinforcing this process is 'de-bureaucratization' by large firms, like BP or ICI, which are trimming their headquarters' staff to eliminate 'unproductive' burdens, a process predicted by Hamilton (1986). Third, the central area is becoming more specialized on very highly profitable services: banking, finance, insurance, management consultancy and advertising. But fourth, vacated or new offices are being occupied by more 'parasitic', wealth-consuming functions like the legal profession. An imbalance is emerging in London's economy between expanding 'parasitic' and contracting 'productive' activities. In 1989, for instance, accountants and solicitors took almost 50% of new net office space in the city region while electronics occupied only 29% (Jones Lang Wootton, 1989a). Finally, completion of the M25 has generated a new wave of rising land values in the outer zone. Land prices for industrial 'nursery units' exceed £500 000 per acre whilst land for high-technology developments has reached £750 000 an acre in the Golden Triangle. This is now driving high-technology manufacturing and associated R & D, which need more land on clean, environmentally-pleasant sites, even further away from London.

These changes raise the question as to how far London can retain its world rank in control functions? The city's ability to attract new business is being weakened by sharpening competition from other world metropolitan centres. London's 'basket' of comparative advantages is being depleted by a rapid deterioration of the quality of life and urban environment combined with rising office costs. According to *South* (February 1990) the average rent per square foot is US$140 in the City of London, US$116 in the West End. Only Tokyo (US$160) is more expensive, while average rents in Manhattan are only US$49. Thus London compares very unfavourably even with New York, as well as with Hong Kong (US$100), Paris (US$70) and Madrid ($47). However, office rents are now rising faster in some NIC cities; in 1989, they rose by 40% in Bangkok, 33% in Hong Kong and 27% in Singapore, while those in London declined by 3%.

The levelling of rents may enhance London's competitiveness, as it results from a sharp increase in supply, particularly in Docklands. Greater London has few serious rivals in this regard. Of new office space occupied in the UK from January to September 1989, 41% was in London and a further 36% in the rest of the South East. According to *Docklands Business World* (12, 1989, 34), 'London as a whole has over 40% of the total new office construction in Europe'. As a result, vacancy rates are rising, but they are still low, reaching 4% in 1989, compared with 11.5% in New York (Jones Lang Wootton, 1989a). London now attracts many overseas investors as leases are long (typically for 25 years) and rent reviews five-yearly: foreigners spent US$4.65 billion in 1989 on such office space, 70% from the Japanese, 7% from the USA and 5% from Scandinavia. Indeed, developers expect to benefit from a general shortage of space in Europe which, they hope, will help fill offices under construction in Central London, Canary Wharf and the western outer metropolitan zone.

Whither London?

As the processes of international integration gather pace, the economic future of the London region depends critically on how well its changing cohort of organizations and firms exploits opportunities, and overcomes constraints to make it the premier entrepot in Europe for business within the global economy. Decision-makers in the private and public sector must respond to a significant shift in the world's productive and technological centres of gravity from the Atlantic to the Pacific, a trend which involves not only rapid development but also closer integration within the Americas and the eastern Pacific Rim to challenge the European Community. Of course, Western Europe may strengthen its position by forging more business ties with Eastern Europe and the USSR, if democracy and stability succeed there, and with Africa, if conflicts can be eliminated there.

London may remain competitive in business links with North America, parts of the eastern Pacific, and Africa, as firms are aided by language and culture. Yet it is in danger of losing some competitiveness in Europe as the economic centre of the continent shifts eastwards and southwards. Cities like Paris, Frankfurt, Munich and Berlin will gain from faster economic growth in mainland Community countries, from financial market deregulation, German unification, and change in Eastern Europe. Madrid could strengthen its role as 'European gateway to Latin America' and, with Barcelona, as a node in 'sunbelt' development.

Changes in, and restructuring by, London's population of organizations in the past 30 years have transformed the metropolis and its region from a mainly materials, to an information, financial and administrative-based transactional economy. It expresses the successful adjustment to dynamic local, national and international forces by enterprises able to trade their comparative advantages in the global arena. With time, as advantages grew or waned 'strategic control' functions in the international economy have passed successively, on balance, from trading to manufacturing firms, then to retailers, and latterly to bankers as each became 'larger' in bargaining terms. Changing cohorts of firms in Central London, and later the city region, reflect these shifts. However, success in economic change depends more on the fusion of clusters of activities, than on dominance by a single productive sub-system. Indeed, the latter exposes urban and regional economies to high risks as firms in a single or similar sectors dominate the character and quality of the local environment, depriving it sooner or later of stimuli to innovation. London's industrial decline was as much a function of the technological backwardness of its manufacturers as of the squalid inner-city or riverside environments they created (and national or local public policies preserved). But it was also linked to the collapse of linked activities such as the seaport.

Currently, London enjoys growth advantages in linked information fields. Mastery of English as a language of global business gives it the edge, but so does the high standard of films. Advert design and mounting depends on close contact between advertising/media firms and hundreds of major UK and foreign manufacturing or service firms in the metropolitan region. The transition to an information society has dubbed London 'a paper metropolis'. If contemporary technology were fully harnessed, it could become 'a paperless metropolis'. Reuters is an example. Sited at the heart of Fleet Street, amid the offices and printing works of leading newspaper firms, Reuters became world famous for the 'latest news' gathered by agents all over the globe who sent their messages to headquarters by telegraph, later by telephone, and more recently by fax. Today, the same office is centre of a global network of 190 000 computers which transmit instantaneous data via satellite on currency-exchange movements by all the world's banks, wherever they are located. So Reuters now 'commands' the global financial information market, an advantage which benefits the City's firms.

If London is to retain international importance, however, its environment will need major overhaul and upgrading. The metropolitan region may continue to substitute tertiary and quaternary employment for manufacturing jobs, but its economic future can only be assured if the population of firms are able to create very substantial net export surpluses, and hence be highly competitive in global markets with internationally-tradeable goods and services. This means making a marked readjustment to high-quality, high value-added products. Yet, as we have seen, London and its outer metropolitan zone are rather dependent on foreign investment and firms. So London and the South East must offer a first class regional environment in quality which is sufficiently competitive internationally to attract and retain UK and foreign investors in the more advanced stages of the production chain: sophisticated manufacturing, R & D, management and information.

Most recent growth in these sectors has occurred in the 'Golden Triangle' because this area meets the environmental criteria required by firms in these activities: excellent access to global air transport, telecommunications, protected countryside, good housing, education, health and relatively easy journeys to work. By contrast, there is 'underdevelopment' in Inner London and the northeast and southeast suburbs and outer metropolitan zone. A squalid urban environment, deteriorating public services, and the social disgrace of homeless people and litter on the streets combine with land and property inflation to drive out or deter business from much of London. Like New York, London lacks the cohesive business community to act with vigour to regenerate jobs and improve the environment. Unlike New York or any other major competing city, London suffers from fragmented city government to promote change. To the northeast, high-technology activities around Cambridge and full operation at London's new Stansted airport, promise a new axis of growth. To the southeast, the Channel Tunnel and new rail link could foster new business growth around Bromley or the M2, M20 and M25 interchanges. But Londoners are likely to have to forego the future prosperity that such growth could bring. The present British government is unwilling to match the vision and investment of the French either by funding a high-speed Channel Tunnel rail-link or by creating new towns with science parks near an intersection of the link with the M25. The price London will pay for the government's narrow 'cost' approach to the metropolitan 'basket' of comparative advantages is that growth will occur across the Channel where, amongst other things, property is cheaper, and Mitterand's visionary Paris will displace London as Europe's control, high-tech and manufacturing metropolis.

Notes

1 The official definition of the South East region is unsatisfactory for delimiting the metropolitan region as it omits parts of East Anglia – like Cambridge or south Suffolk – which have close functional ties with London and its outer metropolitan zone. For simplicity, this chapter uses data for the South East region.
2 A leveraged-buy-out, or LBO, is effectively an acquisition made on credit to profit from splitting up and selling parts of the acquired firm.

5

The City, the 'Big Bang' and Office Development

Derek R. Diamond

The Current Scene

The start of the 1990s finds central London undergoing a building boom of very considerable magnitude, the focus of which is a series of major projects containing massive amounts of new office provision. Substantial growth in financial services and related activities consequent upon 'big bang' deregulation in 1986 is regarded as the primary impetus for the upsurge in new building which either has or is shortly expected to have a very considerable impact on several aspects of London's labour and housing markets. This growth will also create challenges requiring revised public policies as the nature of infrastructure use alters and the spatial structure of the central area adapts in response to significant functional change. This description of a rapidly altering current scene has to be understood in the context of a surprising number of major events in the immediate future which are also capable of generating change, conflict and uncertainty. For instance, 1990 sees the introduction of a new basis for local government finance (the national business rate), 1992 sees the introduction of the single European market and 1993 sees the start of railway services from central London via the newly completed Channel Tunnel.

In addition, these dates occur amidst two important changes for office development in central London – the information technology revolution and new arrangements for the government of London following the abolition of the Greater London Council in 1986. This chapter will clearly show that neither of these are independent of the pervasive workings of the Thatcher revolution. The real consequence of such a period of rapidly changed relationships – ranging, for example, from those between broker and jobber in the Stock Market, through those between developers and local government in land use planning, to the attitude of central government over issues like public expenditure for infrastructure provision or the nature of strategic guidance for the central area of a world city – is the appearance of both new opportunities and new difficulties in a situation seen by most decision-makers as having greater confusion and uncertainty than in the recent past (see Chapter 12). This at least helps explain how the euphoric belief in continued employment growth in financial services in the City in early 1987 (the era of 'golden handcuffs') became the gloomy forecasts of massive redundancies only one year later.

However, important continuities are visible in the current turbulence. Central London (Figure 5.1), an area of approximately 26 square kilometres with between

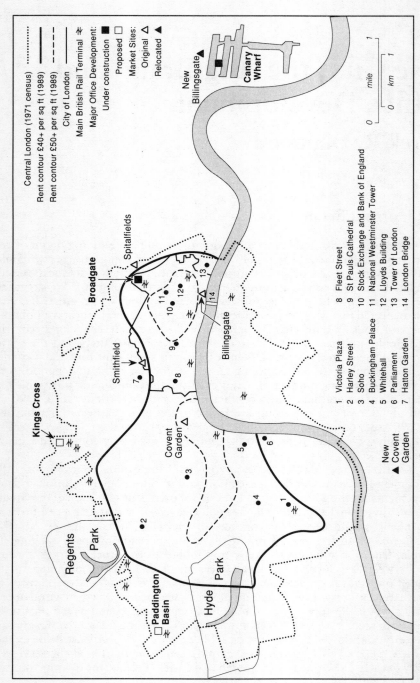

Figure 5.1 Central London

60% and 70% of all London's office space continues to be a location of immense international importance. It is one of the three or four largest concentrations of employment in the world (Hall, 1989, 1) and the dominance of its office-based activities means that it is truly comparable only with Manhattan and central Tokyo. Its much longer history as a world city means that, compared to either New York or Tokyo, the conservation of its archaeological and architectural heritage is a continually live issue as the pressure for change mounts. This is very clearly reflected in the extent of Conservation Areas in both the City of London and in Westminster and, for example, in the continuing argument over the Mansion House Square development in the heart of the City of London and the skyline controls in the latest City of London Local Plan.

Also still visible in the late 1980s is a long established and complex internal spatial structure. This is dominated by the clear dichotomy between the West End, with its shopping and governmental activities, and the City of London, which is centred on the Bank of England and a dense concentration of financial services (Morgan, 1961; Goddard, 1967). Many localities which are famous for a specialized function – such as Whitehall, Harley Street, Soho and Hatton Garden – remain, but some areas have seen significant alterations since the mid-1960s. To cite a few examples, Covent Garden market decentralized in 1974 and its old market was rehabilitated for tourism with retailing, restaurants and a museum, Billingsgate Fish Market closed in 1982, and in 1988 newspaper printing ceased in Fleet Street. Such modifications of long established spatial patterns seem modest in the context of the current developments at Canary Wharf in the Isle of Dogs and the proposals for King's Cross which clearly imply a more massive rearrangement of central London activities than any since the redevelopment era following the Second World War.

Although it will only be clear how great the changes in the form of central London will be when all the space now at the development proposal stage and actually under construction is available for occupation, there is no evidence to suggest a change in the basic principles underlying the functional role of central London. The primary importance of the banking, insurance and related financial services complex seems assured (see Chapter 3), and the interrelated nature of these activities with other major functions – government, international administration, retailing, tourism, the provision of high level business and public services – is as clear as their own inter-relatedness. It is just as likely that central London's role in meeting simultaneously international, national, and metropolitan/regional demands will continue. What is much less clear is the extent to which the operation of these long established principles will be affected by new technological, economic and political circumstances and what the consequences will be for both central London's future importance and its internal spatial structure. One question is increasingly being asked: is the combination of an imperial past, the resultant reservoir of expertise and reputation, and the benefit of the world's time zones (London's Stock Exchange opens as Tokyo closes and remains open until New York has opened) sufficient in the circumstances of the 1990s and beyond to perpetuate London's standing?

This chapter looks first at what is happening to the supply of office space and the requirements of new information and communication technologies before proceeding to examine the changing nature of demand for office space and the competitive threats that exist or may arise which could challenge central London's position. It concludes by attempting to identify the issues whose resolution are crucial for the continued success of central London as a world city centre.

The Building Boom

The late-1980s building boom was almost wholly unexpected, certainly on the actual scale that occurred. In 1982 the Corporation of the City of London noted that employment had fallen by 40 000 in the City between 1971 and 1981 to a figure of 220 000, although it identified a trend to increasing floor space per employee which in 1981 reached 21.4 square metres. Not only did 1982 also see the total number of commuters to central London reach its lowest level since the post-war peak in 1962, but the 1970s had seen a growing trend to office decentralization; much of which was directly or indirectly assisted by the Location of Offices Bureau between 1963 and 1979 (Manners and Morris, 1986). Wood (1986, 67) expressed the typical view of analysts that '. . . although London remains predominant, considerable uncertainty hangs on the future of employment in its private-sector office functions'. Prominent among the reasons for this cautious conventional wisdom were the strength of the office decentralization flows in search of lower rents, better staff and an improved environment, a belief that a greater use of telecommunications would encourage an even stronger dispersal movement, and a reduction in public service employment which, combined with an official civil service dispersal programme, would reduce the demand for space in central London.

However, all the available indices tell a similar story concerning the present dramatic cycle of real estate activity which dates from 1984. In that year there was clear evidence of increasing tenant demand, some of it with specific building requirements, and for the first time since before the 1970s boom (with the brief exception of mid-1978 to mid-1979) the index of annual change in office rents in real terms became positive (Hillier Parker, 1987).

Table 5.1   Changes in Gross Floor Space in the City of London (thousands of sq m)

	Applications	Permitted	Commenced
1977–85 annual average	190 (est)	170	110
1986	570	610	250
1987	805	1180	630
1988	625	(600)	(370)

Source: Corporation of the City of London (1988) and Jones Lang Wootton (1988)

Note: Figures in brackets are for January–June only

However, the upturn in demand was very promptly identified by the development industry, with the data in Table 5.1 revealing the abruptness and scale of response. Between 1977 and the end of 1985 the pipeline of office floor space in the City of London (i.e. that under construction plus that permitted but not yet started) remained stable at about 800 000 square metres. But by June 1988 it had almost trebled to 2.2 million, which, even allowing for the implied replacement of the demolished office space, is a massive quantity in relation to the existing stock in 1988 of approximately 6.0 to 6.2 million square metres (net). According to the review by Jones Lang Wootton (1989) of the five years 1984–88, a total of 2.3 million square metres of office space was given planning permission. The peak in applications for planning permission was reached in mid-1987 with over 840 000 square metres, but it was still at 605 000 at the end of 1988 after permissions had reached a peak in mid-1988. Rent levels rose rapidly after 1985 reaching a record level of £650 per square metre in the City of London in 1988 and £680 per square metre in the West End in June 1989.

Many and varied explanations have been put forward for this office development boom, most of which can be shown to contribute to the total picture, although it is difficult to identify the most significant or novel causes. Setting aside until later the discussion of demand, it becomes clear that the interrelationship of three supply-side conditions is of considerable significance. Suitable locations for what in British terms are very large developments – some unquestionably non-traditional in terms of the existing pattern of location of financial services firms – were identified by developers who, not only used large sites to adopt innovative building designs but, combined this with innovative approaches to development finance and construction techniques. Not only were feasible locations found by realistic entrepreneurs but the economic and financial climate appeared conducive to renewed investment in office property as a result of a combination of factors. Growing confidence in the continuation of economic growth in the British economy, rising inflows of pension money to the financial institutions, steadily falling inflation and growing expectations of the demand consequences arising from 'big bang' were among the more significant of these underlying forces. Nor is it possible to overlook the impact of government policy not only through fiscal regulation but also as an outcome of its urban renewal initiatives. The creation of the London Docklands Development Corporation in 1981 and the subsequent successful attempt to attract substantial development in offices and houses to this area caused a major relaxation of the City of London's attitude to office redevelopment (see the abrupt rise in permissions in Table 5.1).

Also crucial to the speed and scale of a boom born of this combination of key personalities, emerging spatial opportunities and a greatly improved investment climate were two contextual circumstances: information technology and the internationalization of financial services which by changing the nature of demand (see next section) greatly increased the opportunities for profitable office development. Taken as a single entity, because that is the real outcome of their massive interdependencies, these conditions and circumstances constituted an explosive mixture, the full effects of which will remain unclear for some time to come. Nor in any full analysis is it possible to ignore continuing problems – the perennial weakness of British manufacturing industry, the massive remaining stock of out-of-date office buildings, and infrastructure obsolescence – although in the late 1980s these were largely obscured by a dramatic office building boom, whose typical product looks quite unlike the classic high-rise model of the 1960s and 1970s, and which is creating a new geography of employment in a greatly enlarged central London.

Unlike the NatWest Tower, London's tallest office building, and unlike the glistening high technology of the Lloyd's building, Broadgate (which is beside Liverpool Street Station and was commenced in 1986) typifies the new wave. It involves 14 buildings on a 12 hectare site with a total provision of some 370 000 square metres of advanced design office space and is the largest single development in the City of London since the Great Fire of 1666. Undertaken by the Rosehaugh and the Stanhope property companies led by Godfrey Bradman and Stuart Lipton, the new buildings are only eight-to-ten stories high but their post-modern style cannot disguise their functional features – massive uninterrupted floor space (ideal for the new style trading floors of the equity or foreign exchange dealer) with controlled environments to extract the heat from the computing equipment, and between the floors facilities for huge quantities of wiring. More floor space per employee was required not only for the rising standards evidenced in the atria designs and in such external amenities as the ice-rink or shopping arcade, but above all to accommodate new information technology.

Deregulation of London's financial services by agreement between the Secretary of State for Trade and Industry and the head of the Stock Exchange (implemented on the 27th October 1986) had its intended effect of encouraging competition as well as foreign participation in the Stock Exchange at the same time as it allowed for all stock exchange transactions to be fully computerized. This agreement, to open up London's financial markets was reached only after the Office of Fair Trading threatened to take the Stock Exchange to court under the Restrictive Practices Act. Quite appropriately on this occasion, the combination of new rules and new tools at the heart of one of the most conservative elements of the British establishment, was referred to in the media as the 'big bang'. The most immediate and dramatic evidence of the new situation was the empty and abandoned trading floor of the Stock Exchange as market-makers increasingly switched to operate from VDUs in their City offices. The fundamental changes in the structure of financial services engendered largely, but not wholly, by the big bang and the reformed regulatory system were compounded in a complex manner with remarkable advances in information technology that had occurred in the previous 20 years. It is often overlooked that the introduction of these technologies in New York some years earlier provided a clear demonstration of their reliability and effectiveness. Indeed, the developer of Broadgate is on record as saying that his firm undertook an evaluation of the new dealing rooms in New York before commencing this development.

The combined effect of the radical organizational and technological changes was that almost overnight much of the existing office stock in the core of the City of London was rendered obsolete in the sense that it was no longer efficient for the larger, multi-functional financial service firms which were emerging. The 5000 square metre dealing room was merely the most extreme case of a widespread problem created by the sudden change in ways of working of, particularly, financial services but also many other producer services such as accountants, management consultants and software providers. In these circumstances it was not surprising that a traditional location within a short walk of the Stock Exchange should be given a lesser priority when accommodation is available elsewhere which offers an efficient operating environment. This thesis will be fully tested when Canary Wharf (phase 1) within the Enterprise Zone on the Isle of Dogs becomes available for occupation in 1991. The application for planning permission for a further 560 000 square metres of offices in the proposed King's Cross development by London Regeneration Consortium PLC., appears, however, to confirm that such non-traditional locations are now regarded as viable.

This proposed massive northward and eastward extension of the traditional office zone is not just the result of suitable sites providing a Hobson's choice for new and expanding financial service firms, though this is a real enough pressure in the 1988–91 period. It is also a recognition of the way in which the application of information technology has loosened traditional location requirements. In this process of widening the area of viable operation it has also increased the likelihood of stronger competitive challenges from other cities, thus focusing attention clearly on the critical role of operating costs. If Olympia and York, the Canadian property company which is constructing Canary Wharf, achieve their stated aim to offer office accommodation with twice as good an environment at half the rent of that in the core of the City of London (circa £230 per square metre), then clearly a serious challenge to the traditional pattern is underway.

In this context the speech of the Governor of the Bank of England on 17th October 1985, contained an important shift in policy:

> The City [of London] will certainly provide some of it, perhaps with some further development. I think it doubtful, however, whether it can, or indeed would want, to provide it all. . . . In this situation developers, their clients and the planners are looking at the alternatives. This naturally raises the question of our attitude at the Bank of England. Traditionally we have preferred the banking community to locate within easy reach of each other and ourselves. In practice this has tended to mean clustering within the 'square mile', although several large banks have already moved away. Perhaps the shift in emphasis from personal to electronic communication has altered the balance of the argument somewhat. If we are to achieve our goal of keeping London as one of the three major international markets in financial services, the people who provide them must have the physical facilities they need. For that reason financial institutions should choose for themselves where to locate in the light of commercial considerations. It is not our intention to stand in the way of their judgment. We hope, however, that the City itself will retain its cohesive character. There is space within reasonable reach of the City to meet an overspill of demand. In short, a combination of imaginative adaptation by the City and complementary development in adjacent parts of London, particularly perhaps those whose traditional livelihood has largely vanished, would seem an appropriate evolution.

Several other factors were operating at the same time to facilitate change in attitudes that enabled non-traditional sites to gain acceptability. Two of these which were notably compatible were of particular importance; one was British and the other overseas in origin. It would be difficult to overestimate the significance of the ethos and policies of the Thatcher government. Beginning in 1979 with the abolition of foreign exchange controls, of office development permits and of the Location of Offices Bureau, the government's consistent approach to greater competition embodied the 1982 agreement with the Stock Exchange, the privatization of nationalized industries and the setting up of the first Enterprise Zone in London's Docklands some six kilometres from the Stock Exchange, all of which had direct effects on the operation of financial services. The impact was greatest on those that were involved or could readily become involved in international activities such as overseas investment, foreign exchange transactions, and sovereign debt management. These were almost solely located in the City of London. One notable consequence was the growing presence in the square mile of non-British financial services firms; most particularly Japanese and United States banks and their associated investment arms (King, 1990). However the rapidly increasing foreign presence (see next section) was not limited to banking and finance but extended to participation in major office development schemes (e.g. Olympia and York) and to the construction boom itself.

If it is true that the rise in the volume of shares traded, the record levels of the equity indexes in London, Tokyo and New York, and the enormous pension fund activity of the early 1980s (Plender, 1982), were a happy coincidence, then the attitude of British Railways, the London Docklands Development Corporation, most local planning authorities, and the Corporation of the City of London itself were clearly a response to the enterprise economy with its freer markets that was a central aim of the government. It is not just coincidence that Broadgate, Canary Wharf, King's Cross, and many other proposals have come to life at just this point in time. Unused land, which was surplus to current operational use had existed as docklands, railway goods yards, redundant markets etc., for many years previously.

Although each of three major sites discussed are very large office developments by any definition – Broadgate 370 000 square metres, Canary Wharf 930 000 square metres, King's Cross 560 000 square metres, the total amount of new floor space is even more impressive. If all the existing proposals which can reasonably be labelled central London are completed by the mid-1990s, this would be equivalent to building half of the City all over again. Of course a considerable, if uncertain, percentage of this is replacement and not all applications will be granted. Nevertheless, current plans clearly indicate a modified spatial structure is emerging in central London.

The vast majority of the new provision is found in three broad bands. The fringe of the City of London is typified by Broadgate but includes the south bank of the Thames (e.g. London Bridge City) as well as the west Holborn Viaduct Station and the redevelopment of Fleet Street (which is already home to more than 30 sites each with a minimum of 90 000 square metres). A second concentration, more in the planning phase than under construction, is the remainder of the fringe of central London, that is its north, west and southern sides where railway station schemes loom large, as at King's Cross and Paddington. Third is Docklands, with Canary Wharf as its centrepiece. Though no-one expects the growth in financial services alone to occupy all this space, their role as the motor for the expansion is now widely accepted. This raises the question of the future of their relationship to other producer services which occupy office space in the heart of central London. To maintain anything like the convenience provided by the 1980s pattern of proximity implies very considerable locational adjustments or greatly improved accessibility within central London, if not both.

The Nature of Demand

The rise in office rents to record levels in 1989 provides the most direct evidence of the strength of demand in recent years. Since 1977 the most rapid increase has been in the West End, where currently rents are almost identical with those in the City which is in marked contrast with the peak of the previous boom in early 1974 when, '. . . prime rents reached £22 per square foot in the City and £12 per square foot in the West End' (Manners and Morris, 1986). As Figure 5.2 vividly shows, the City of London and West End are merging into a single office market under this pressure of demand.

It was widely reported in 1988 that many international tenants are now more willing to consider all of central London as a suitable location, rather than limiting their search to the City of London. The choice of Victoria by Salomon Brothers for their 300 dealer trading room in 1979 now appears as a forerunner of present trends rather than as an exceptional case, illustrating the impact of the shortage of space (in particular, the shortage of suitable space for larger, international firms with enormous investments in information technology). The stronger emergence after deregulation of these major players, who are capable of dealing round the clock, has changed the nature of the securities industry. The comprehensive financial services of these companies (which include commercial banking, merchant banking, fund management, market making, stockbroking) are increasingly available in all major financial centres throughout the world.

In the period 1984–88 Jones Lang Wootton have estimated that the total take-up of floorspace within the City of London and its fringes amounted to 2.4 million square metres; which is the equivalent of about one-third of the original stock. In 1984, 335 000 square metres were acquired for occupation. This was an increase of 30% on the previous year, and from 1985 to the end of 1988 it

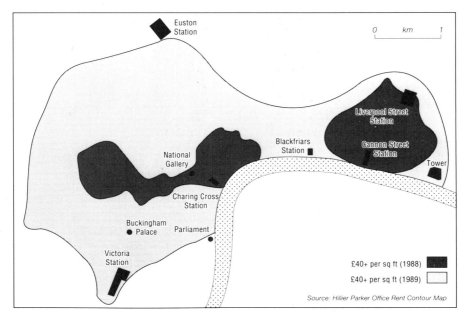

Figure 5.2 map labels:

Euston Station

0 km 1

Liverpool Street Station

Blackfriars Station

Cannon Street Station

National Gallery

Tower

Charing Cross Station

Buckingham Palace   Parliament

Victoria Station

£40+ per sq ft (1988)

£40+ per sq ft (1989)

Source: Hillier Parker Office Rent Contour Map

Figure 5.2 Expansion of the prime rent zone 1988–89

increased by a further 60% to run at almost 560 000 square metres per year. At the end of 1988 one major assessment of demand estimated the current need at 570 000 square metres, clearly indicating the continuing strength of demand for suitable accommodation. Remarkably, more than 70% of total take-up in the period 1984–88 has comprised new rather than second-hand floor space. This provides a further indication of the need for large uninterrupted floors, for modern servicing with ample vertical ducts and for raised flooring. In fact, much of this take-up (over 80% in 1986–88) has been pre-let; that is, it has attracted a tenant before being available for occupation, so great was the scarcity of suitable accommodation at the time of big bang.

In all, 46% of new take-up between 1984 and 1988 in units larger than 9300 square metres was in the banking and securities sector. Here, many firms sought to centralize their enlarged operations under a single roof. In the five years 1984–88 banking and securities accounted for 57% of the total take-up of new space (in all over 930 000 square metres) with the slightly larger share going to overseas companies compared to British firms. A significant amount of new space has been occupied by accountants and solicitors (notably in 1988), followed by such business sectors as information technology and marketing. The traditional City sectors, such as insurance, shipping and commodities took up 37 000 square metres of new City of London floor-space in 1984 but in subsequent years averaged a more modest 18 500 square metres.

There is a close correspondence between these data on take-up by types of activity and growth in employment. Rajan's (1988) survey of what he calls 'the six heartland industries' of the City (Table 5.2), which were selected for '. . . their interrelated leading edge features that give the City a unique role in both the labour market and the wider society', and which account for approximately 65% of all service jobs in the City, expanded between 1984 and 1987 by 51 000; which is equivalent to eight times the national average (see Chapter

Table 5.2   Employment in the City: 1984–87 and 1987–92

	Actual				Forecast	
	City employ-ment (000s) 1984	1987	Annual % change 1984–87 City	GB	City employ-ment (000s) 1992	Annual % change 1987–92
Banking	93	114	7.5	1.5	125	2.0
Other financial	38	43	4.5	5.0	48	2.5
Securities dealing	18	28	18.5	17.5	25	2.0
Insurance	40	45	4.0	4.0	50	2.0
Accountancy and management consultants	20	25	8.5	6.0	36	9.0
Software services	13	18	13.0	11.5	26	10.0

Source: Rajan (1988)

Note: 'City' is an abbreviation for the City of London

3). Banking and securities accounted for 61% of this growth but it is the securities and software services rates of growth that were twice those of the other heartland industries. Overall these employment changes have tended to increase further both the concentration in central London of financial services and their interdependency.

A much commented on underlying influence behind growth in the financial sector throughout the 1980s has been the foreign banking sector. According to the London Chamber of Commerce (1989), employment in foreign banks and security houses increased from 21 000 in 1975 to 62 000 in 1987 and half the growth in the financial sector (excluding insurance) in the City of London in the 1980s was accounted for by foreign banks (although some of this was the result of the acquisition of British-owned firms). London accounts for around one-fifth of total world international banking which in 1989 gave it a share equal to that of Japan and twice that of the USA (Bank of England, 1989). This is perhaps an obvious measure of internationalization in the financial sector but it greatly underestimates the genuinely international nature of present-day financial services. It has recently been estimated that 40% of the earnings of the FTSE 100 Index came from abroad, despite their label as domestic securities. The close relationship between the expansion of demand for accommodation in the City and internationalization is well identified in a recent Bank of England report which states that:

> A market feature of recent years has been the expansion of Japanese banks' international role. Since 1982 they have been the largest national banking group, as measured by balance sheet size, in the world. London is the largest centre for Japanese banks' international business outside Japan itself, with some 26% of Japanese banks' international assets booked here (at end-1988), bringing their share of all international lending out of London to 36%, as against 13% in 1975. (Bank of England, 1989, 518)

Towards the end of 1989 evidence emerged that the rate of demand for office space was slowing and it was possible to forecast that, given the large supply currently in the pipeline, some surplus capacity would probably appear in the 1990–92 period. In line with the slow-down in the UK economy, which was heralded by the introduction of a bank rate of 15% in October 1989, take-up eased from its high levels in 1987 and 1988. Rental growth has slowed and vacancy rates have increased by two or three times from their previously low levels. While the overall central London office market vacancy rate is expected to rise to 6–7% in 1990 (from 2.5% in 1988), which is a rate that is below that of New York, wide variations are expected within the central area. The City of London is expected to have vacancy rates of about 10% while the West End is expected to have only 3%. At the other extreme, Docklands is predicted to face a 40–50% vacancy rate as it struggles to establish itself as a convenient alternative to the City of London but for the first part of the 1990s this area has only limited accessibility to the rest of central London. These differences reflect the emerging spatial structure of the newly enlarged city centre. However, this structure can be expected to alter considerably by the mid-1990s (see next section).

In contrast to this somewhat depressed domestic scene, overseas investment has remained at a high level and domestic financial institutions have been able to sell into a strong market for the largest and most prominent buildings. Overseas investors, notably from Japan (22%), Scandinavia (12.5%) and the USA (3.5%), accounted for 50% of the funds placed into central London offices during 1989. This continued strength of overseas investment suggests a reasonably optimistic view of the longer term prospect, and is one that recognizes the notoriously cyclical nature of conditions in the central London office market.

While it is unquestionably true that the immediate cause of the explosive growth in City jobs, with its consequential need for additional accommodation, was the deregulation of financial services, this explanation overlooks the important role that major structural trends have played, and will continue to play. Among these structural trends are considerations that, '... inspired deregulation, others that made it inevitable, and others that complemented it' (Rajan, 1988). All have longer term implications for the future of employment in the City. These important structural features include such trends as the internationalization of business, organizational changes in producer services, and a complex set of changes that includes product innovation, firm diversification, and new modes of production (increasing knowledge intensity combined with rapidly developing information technology applications). There is not the space in this discussion to elaborate on this complicated situation, in which '... concentration (and associated with this acquisition and merger), diversification, deregulation, internationalization and technological development' (Howells and Green, 1986, 177), is having a fundamental impact on the geography of producer services everywhere (Price and Blair, 1989). However, the changes currently being wrought by the combined effect of these powerful trends have the potential to alter radically both the kind of services that are required and the way they are used.

Indeed, it is the very success of London's financial institutions in creating a diverse and deeply interrelated set of tradeable services that makes the role of the structural trends so significant. A recent Bank of England report (1989) on London's role as a diversified global financial centre identified the following world roles: international eurocurrency business; eurobond transactions; foreign exchange; fund management; corporate financial advice; equity trading; insurance and trading futures and options. New value-added services emerging from such a rich product-mix, and from their interaction with information technology, will be the basis for the future demand for additional office space in central London.

Towards 2000

It is true that the service function of the City predates the rise of manufacturing and most of its services have always relied more on the existence of other services than they have on manufacturing industry (Goddard, 1968). It is the very nature of current trends that these historical assets can become liabilities if imagination and flexibility is not shown by those with the inheritance. The trend towards knowledge-based, skills-intensive occupations is a crucial challenge to the City if it wishes to remain '. . . at the leading-edge of developments heralding a knowledge society' (Rajan, 1988, 9). To meet this challenge raises questions of training and of the spatial distribution of labour. These are equally important as those concerning the physical provision of offices.

The changes wrought in the supply of office accommodation in central London in the late 1980s and in prospect for the 1990s, together with similarly important structural changes in the demand for office space, have laid the foundations for significant changes in the extent and structure of central London. Indeed, as measured by institutional investment, even the national significance of central London has increased, with commitments to offices increasing from 51.5% to 64.1% of the national total between 1980 and 1987, while the centre's retail share fell from 8.8% to 5.6% (Nabarro, 1989). However, the multiple and complex causation that lies behind the recent boom in office space makes forecasting future trends immensely difficult. Nevertheless, two of the new key elements in the future pattern of central London are now clear.

First, the internal differentiation of the central London office market is being massively increased. A process, remarkably similar to that which affected retailing a decade ago (see Chapter 7), is producing marked segmentation. Widely different accommodation needs, associated with specific technological and organizational requirements, are creating a wide range of provision that is as diverse as the contrast between corner shop and hypermarket. No longer will it therefore be useful to generalize trends in the office market on the basis of a notional 500 or 1000 square metre suite. In terms of current trends much of the existing office stock is obsolete. According to one estimate, central London has some 14 million square metres of office space built before 1980, much of which is unsuitable for the electronic office, for rooms are too small, they are the wrong shape and they have wholly inadequate ventilation.

Recent research has confirmed previous findings that the most important factor behind firms relocating outside central London is the reduction of operating costs. One-in-three firms has reported that operating cost savings were the main reason for decentralizing, while a further 25% gave the centralization of their activities and the upgrading of accommodation as the main reason. A better match of supply with demand in a segmented market with its varied and distinctive sub-markets is clearly a precondition of the continued success of the central London office market.

Thus, given the crucial importance of being able to offer an efficient operating environment both to domestic and to overseas firms in the increasingly competitive conditions of the 1990s, the opportunities provided by the major peripheral development sites can be seen to be essential to the long-term success of central London. It is almost only in these locations – Canary Wharf, Kings Cross, Broadgate and Paddington – that development on a scale sufficient to create the missing elements in central London's office provision can be achieved without massive change to the existing urban fabric. While important questions of scale and phasing of these projects remain to be settled, their long-term strategic value is not in doubt and the consequent re-shuffling of firms within central London can be expected to occur over a considerable period of time.

Arising out of the market-driven reallocation of office provision emerges a new geography of employment in an expanded central London. It is instructive to note that the expected two million square metres of new office provision anticipated by mid-1990 has the capacity to provide for at least 100 000 employees. The most obvious feature of this is the increase in employment density in a series of locations in what is now referred to as the fringe of central London. This spatial arrangement, which can only be fully achieved with the agreement of the local planning authorities (see Chapter 11), has both advantages and disadvantages.

In general this fringe employment growth will be more accessible to residential London (especially inner city London) and, provided attention is paid to the provision of social housing in conjunction with major schemes, a more convenient journey-to-work could be achieved for many. However, these somewhat scattered locations will also generate a great demand for interaction between them, with the rest of central London and, in particular, with the City of London. As the central area expands so it brings with it a demand not only to extend the level of accessibility that now exists to its new additions, but to make their travel connections to the rest of Britain and to the rest of the world at least as efficient as those of its main competitors. The 1990 (January) survey of world rental levels, carried out by the estate agents Richard Ellis, ranks office centres in terms of total occupation cost (rent plus service charges and property taxes, if any). Using the City of London as the base, with an index score of 100, the comparative index scores obtained were as follows: Tokyo is the most expensive (142), followed by City of London (100), London's West End (96), Hong Kong (73), Paris (56), mid-town New York (45), Milan (41), Madrid (41), Sydney (37), Singapore (37), Stockholm (36), Frankfurt (33) and downtown New York (33). Of course, the apparently expensive nature of central London's offices is partly a reflection of its importance as a world financial centre but it also indicates that there are cheaper city centres in the same time zone in Europe. If office rent is not the whole story because inter-firm and inter-sector relationships are also extremely important for success, then the necessary physical infrastructure to allow such relationships to happen efficiently is also important for central London's continued success.

It is therefore an important paradox that just as some of the economic and spatial parameters of central London's future have become visible, and their implications for the public sector have become understandable, there has been a fundamental disruption in all the political and administrative arrangements that facilitate collaboration between the public and private sectors. Such mechanisms are of particular long-term importance for establishing a sense of direction and in providing the necessary continuity to make large investments in infrastructure worthwhile.

Precisely which combination of factors is responsible for the dismaying fragmentation of transportation infrastructure improvements in central London either under construction or under consideration can be debated, but the piecemeal nature of the present situation is beyond dispute (see Chapter 8). The agreement to extend London Underground's Jubilee line from the West End (Green Park) to Canary Wharf and Stratford in the east via Waterloo, together with the extension of the Docklands Light Railway to Bank and to the City Airport are important but are not easily related to a clear strategy for an expanded central area. Much the same will almost certainly be said of the eventual decisions on how to connect central London to the Channel Tunnel high speed trains to Europe, and any project that stems from the central London Rail Study (Department of Transport 1989b). This study was commissioned to consider alternative improvements to London's rail infrastructure in face of the 200 000 additional daily passenger arrivals on London's underground since 1982.

There is no doubt that it is difficult to understand what infrastructure central London needs by 2000 to perform effectively its role as simultaneously the metropolitan centre of South East England, the national capital and a world city. But without such a perspective there is no answer to such key questions as, what balance should be struck between infrastructure that aids the international role of London by improving access to external connections as against giving priority to the relief of commuter congestion (Chapter 12)? Indeed without a strategic overview it is not even possible to explore whether or not there is an option that might serve both objectives.

Nor is the transport question the only one to raise serious doubts about the effectiveness of planning for the future of central London. Current issues of real importance that exist entirely within the control of the British government include how to resolve serious labour shortages in the public sector (school teachers, nurses, social workers, etc.), how to assist local government management to undertake effectively reforms in finance, education, housing and the competitive tendering of services, and how to devise better machinery for evaluating and deciding on a vision for the future of central London.

Outside the UK it is Europe that appears most likely to produce a major new impact. Substantial economic gains from the real integration of European financial service markets are expected following 1992 because of '. . . the pivotal role played by such services in catalysing the economy as a whole' (Cecchini, 1988, 37). Three interlocking outcomes are anticipated: '. . . a surge in the competitivity of the sector itself, a knock-on boost to all businesses using its increasingly efficient services and a new and positive influence on the conduct of macro-economic policy in the EC' (Cecchini, 1988, 37). In Cecchini's (1987) report it was found that the UK had a substantial price advantage in six out of a representative sample of 15 financial products investigated, namely in institutional equity transactions, life insurance, mortgages, motor insurance, travellers' cheques and public liability cover (in decreasing order of price advantage). This complexity will clearly cause differential benefits among the main financial centres such as London, Paris, Amsterdam, Frankfurt, Milan, and possibly even Zurich. Nor does it make sense in the light of the history of massive interdependence among financial institutions to ignore the existence of new intergovernmental financial agencies (such as the proposed central bank for the European Community).

Conclusion

In the introduction of this chapter the question raised was, is the history that created central London sufficient to command its future? The answer is clearly no. As the Bank of England has recently implied, a successful future will not just happen, even though:

> . . . the wealth of experience and expertise available in London, the advantages of an established centre displaying the full range of financial and ancillary services and the determination to improve systems where necessary suggest that London should be well placed to meet the competitive challenge. But neither the authorities nor firms operating here can afford to be complacent (Bank of England, 1989, 528).

If sufficient decline in the standing of central London as a world centre is a real possibility, albeit the result of many different types of threat, then the need for an informed debate leading to a positive response is urgent. In the early months of 1990 several organizations, both public (e.g. Labour Party) and private (e.g. the estate agents Hillier Parker), have indicated a serious concern about the ineffectiveness of existing arrangements. The variety of models being suggested, which range from a Minister for London through an appointed Planning Commission for central London, to reinventing the Greater London Council, indicate the difficulties that exist. However, some of the essential features required are clear either in the lessons of the past or in current difficulties.

First of all, any interventionist public policy, even of the mildest kind, which is faced with the kind of complexities discussed in this chapter must be well informed and be able to take a medium to long-term view (of 20 or more years). Reviewing the experience of office location policy from 1963 to 1980 Manners and Morris (1986, 156) concluded that: '. . . an arrangement for measuring and regularly monitoring the scale and geography of employment in office activities is therefore of paramount importance and should be instituted at the earliest opportunity'. This is a task that not only needs the full co-operation of the private sector if it is to succeed but which, because of what John Goddard has called the convergence of the within workplace technology of computing with the between workplace technology of telecommunications, will require new concepts and new types of data. Among the now possible major changes in organizational structure that could significantly affect the central London office market would, for example, be the establishment of a government data network that would lead to major changes in the location of civil service jobs.

Second, there must be a focus on the integration of major infrastructure improvements and their development potential. A wide range of infrastructure types are relevant since inter-city competition for international firms includes what have come to be called the quality-of-life rating that such cities have. It is currently fashionable to claim that the location of the rapidly expanding producer-services sector is primarily determined by the need to recruit and retain expensive staff who have strong views not only about the internal office environments in which they work but also about the external environment. The extension of the Royal Opera House, the new wing of the National Gallery and the Tate Gallery extension, along with the new British Library (£300m) are illustrative of this theme in London, if rather less dramatically than the bicentennial equivalents in Paris. It remains to be seen if the economic effect of public expenditure on culture can be assessed in some way but it is already clear that this aspect should not be ignored (for a discussion of this in the context of the USA, see Whitt, 1987). But it is transport and communications infrastructure that is both crucial and currently under the greatest pressure (Chapter 8). It is the essence of world city centres that they are infrastructure intensive.

Third, there must be successful working relationships between central government, local government, infrastructure providers and managers and representatives of the business community. It is an irony that the very best government whose deregulation policies made it possible to seize the opportunities emerging from internationalization should also be responsible for the failure to coordinate public investment in infrastructure with private sector development initiatives. The consequential piecemeal and fragmented programme makes life more difficult than it need be for all who use central London; local citizens and overseas visitors alike. As people and societies get richer they demand ever more a better environment and the only way that they can get it is through public action that coordinates diverse interests, matches progress in London with its serious competitors and provides a view of the future for central London.

It is deeply worrying that, at the very point in time when the intricate web of internal relationships which reflect regional and global networks is being restructured (under the influence of powerful trends arising from new technologies and increased competition), there is no suitable machinery for such an essential task. In its absence there are those who see a bleak future for central London, as economic decline in the UK as a whole, and greater deregulation overseas allows the repatriation of international finance to competitor cities, as well as those who rest their case for a buoyant future on the enormous lead and expertise already established.

London emerged as a world city through Britain's international political role and extended it vastly through the application of special skills in financial trading. It now has to face an increasingly competitive situation without the former and with clear difficulties to maintaining the quality of the latter. In this context it seems somewhat unlikely that the existing arrangements will succeed in providing the excess of worthwhile combinations that is the essence of the successful world city.

6

Gender Divisions of Labour
Simon Duncan

Introduction: Familiar and Unfamiliar Geographies of London

Compare Figure 6.1 and Figure 6.2, which both map social distribution by London borough. The first map shows the percentage of 'economically active' people who held managerial and professional jobs in 1981, and this is presented as an index of geographical variations in class structure within London. (This map, and most of those that follow, also give comparative information for Britain as a whole and for the other metropolitan areas in England.) A familiar geographical pattern emerges, with contrasts between inner and outer London, and between east and west London. The second map, Figure 6.2, shows the distribution of full-time, but unpaid domestic workers as a proportion of all 'working women' (defined here as this group together with the 'economically active' in paid work, see Appendix). This map presents a substantially different, and much less familiar, social geography. New groups of boroughs are formed with seemingly little reference to class structure. Quintessential East End boroughs of low social status, like Barking and Newham, are now in the same category as notoriously high social status, suburban boroughs like Bromley or Sutton. Similarly, the boroughs with the lowest proportion of full-time domestic workers include high-status boroughs in west-central London, like Kensington and Chelsea or Camden, and low status boroughs such as Lambeth.

Both maps show that there is a significant spatial variation in the division of labour and the way in which this interacts with housing markets. However, Figure 6.1 focuses on class divisions in paid labour which centre around formal workplaces and are allocated by the labour market. Figure 6.2, in contrast, looks at gender divisions of domestic labour, work which is largely carried out in private households and is allocated following negotiations between household members. The first map of professional and managerial workers refers to people of both sexes – although men comprise nearly 80% of this group. The second map of unpaid, but full-time, domestic workers refers only to women. This is because there are virtually no full-time 'househusbands', at least as far as the British census is concerned (see Appendix). It would appear, then, that the geography of gender is substantially different from the geography of class.

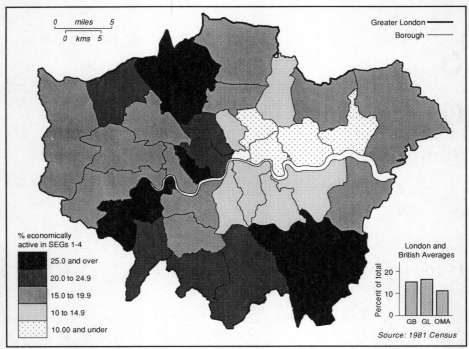

Figure 6.1   Professional and managerial socio-economic groups, 1981

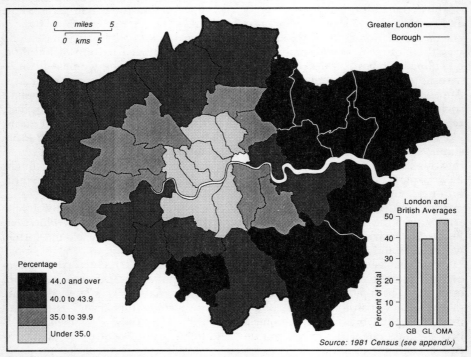

Figure 6.2   Working women in full-time domestic work, 1981

The spatial patterning of class structure within London shown in Figure 6.1 is both well-known and relatively enduring. It can be extended or broken down by using different or more sophisticated indices than those used here, but the same broad pattern persists (e.g. Congden, 1986). Similarly, this pattern correlates with other measures of privilege and status, such as levels of car ownership or owner-occupancy. In both cases, what deviations occur can generally be explained by reference to qualifying, but essentially contingent, factors. For instance, levels of car ownership are lower than one might expect in the high status boroughs of west-inner London because the utility of owning a car is lower in central London, while the over-representation of single, often transient, young people there reduces owner-occupancy rates. This geography of London's class structure can be explained by general social processes like suburbanization, gentrification and labour market specialization, in combination with particular historical factors like the original location of sites of political power in the West End or the necessary location of docks in the East End. As such, occupational and class distributions are a basic feature of London's social, political and economic geography.

Occupational differences are a basic component of capitalist divisions of labour. However, a major part of the labour carried out in capitalist societies is domestic work undertaken outside formal labour markets; capitalist markets do not in themselves produce and maintain socialized workers and consumers, nor do they play more than a very minor role in bringing up children or caring for other dependants. Although the British state is vitally concerned with such matters, it too leaves the vast bulk of such work to be undertaken by domestic workers who are unpaid in the sense of a formal wage or labour exchange. In the famous words of Beveridge, in his 1942 report which set guidelines for the British welfare state:

> The great majority of married women must be regarded as occupied on work which is vital enough though unpaid, without which their husbands could not do their paid work and without which the nation could not continue. (quoted in Mackenzie and Rose, 1983)

This domestic labour is pre-eminently women's work, both quantitatively and qualitatively. Quantitatively, surveys suggest that the vast bulk of such work is performed by women (e.g. Gershunny, 1983; Witherspoon, 1985, 1988; Ashford, 1987). Domestic work is so ingrained as women's work that even women in full-time paid work, with part-time or unemployed husbands, still do most of the housework (Laite and Halfpenny, 1987). Men, of course, undertake such work from time to time, and may even dominate a few, specialized tasks like tending cars or mending household appliances. Some without access to the labour of wives or mothers must even look after themselves. Only rarely, however, do men take on domestic work for others or care for dependants, still less do they take the responsibility for planning and executing this work. Hence, in 1981, there were at the very most 14 000 men below retirement age in London employed in full-time domestic work (the true figure is probably much less, see Appendix). But 928 000 women were in this position. This situation accords well with survey evidence showing an overwhelming preference, in Britain, for an ideal household of a breadwinner father and a homemaker mother (Ashford, 1987). In practice 33% of partnered dual households in 1983 adopted a 'compromise' position, with women in part-time employment, especially when children were at school. In 27% of dual households both partners were in full-time work, with 35% in the 'ideal' role of full-time male breadwinner and full-time female homemaker (Martin and Roberts, 1984).

Domestic work is accorded low status (with the exception of those few jobs mostly carried out by men) and is usually badly rewarded in monetary terms. The value of domestic work is not usually recompensed at market rates, and it was only in 1970 that women in Britain gained an automatic legal claim to the material wealth accumulated during marriage. Domestic work, and hence women's work, is also crucial in defining social roles. Thus, what sort of paid work women and men are able to do – or are thought capable of doing – is partly defined by this gendered relationship to domestic work. The division of domestic labour will, therefore, play a direct role in determining people's life quality and experience, as well as an indirect one in influencing areas of life like access to paid work.

Class differences in the division of labour show a spatial logic – 'spatial divisions of labour' (Massey, 1984). Capitalist work functions are spatially separated, so that different concentrations of occupational groups occur in various areas. This spatial division also interrelates with socially segregating housing market processes to produce distributions like those shown in Figure 6.1. These divisions of labour are also gendered; for mechanisms of occupational sextyping are remarkably rigid and enduring, so that people are generally allocated to 'men's jobs' and 'women's jobs', respectively. This gendering of labour can thereby explain different historical and spatial patterns in male and female employment (cf. Walby, 1987).

At first sight we would perhaps not expect an equivalent 'spatial division' of domestic labour – for in modern Britain households normally centre around a male–female couple living in the *same* dwelling. There are two important reasons why this expectation could be incorrect. First of all, there is an increasing diversity of households – in 1986 only 63% of British households consisted of this supposed normal type. In other types of households women may be freed of domestic work for others and men may have to look after themselves. Secondly, it must be remembered that domestic work is allocated by negotiation between household members. Usually, men have greater power *vis-à-vis* women in these negotiations, but there is good reason to believe that the degree of this varies even within households of the same type – depending for instance on whether women have their own full-time paid job or not (see Laite and Halfpenny, 1987; Morris, 1989). And we would expect both these factors, household differentiation and differences internal to the household, to be spatially patterned. Certainly Figure 6.2 suggests that in London at least this variation can be quite significant (cf. Halford, 1989).

Maps of variations in class structure are commonplace in accounts of the geography of London, but as far as I know, Figure 6.2 on domestic workers has not been presented before. The point of this chapter is to correct this deficiency in previous geographies of London and examine gender divisions of labour. In doing so I will build on previous work on spatial variations in the gender division of labour (e.g. McDowell and Massey, 1984; Bowlby *et al.* 1986; Walby, 1986; Bagguley and Walby, 1989). So far, this work has not been extended to the sort of ecological mapping I undertake here (although see Hodgson, 1984 for a pioneering statistical account of women living in London).

I carry out this task in three sections. In the next section, the relationship between class and gender is examined, both empirically for London and theoretically in terms of the relationship between capitalism and patriarchy. This relationship, I conclude, is spatially patterned in important ways. Then, using this information as a starting point, the next section examines the generative processes that create this spatial patterning. This section looks particularly at differences in male/female social mobility in labour and housing markets. I end with some comments about future research in the context of historical change and continuity.

The Geography of Gender Roles

Class and gender – contrasts and combinations

Class and gender relations do not, in practice, normally operate apart. Class is gendered, and gender is stratified (cf. Crompton and Mann, 1986). Take the map of class structure presented in Figure 6.1. There are far more males in this professional and managerial group, 446 000 compared to only 129 000 females. Mapping the percentage distribution of *male* professionals and managers only (Figure 6.3), the most striking pattern is the higher relative presence of this occupational group – only one borough (Newham) now scores less than 10% while 11 score over 25%. The spatial pattern also changes, for the suburban concentration of this occupational group is stronger (with five suburban boroughs scoring over 30%). Contrast this pattern with the picture for female professionals and managers (Figure 6.4). Immediately obvious is the far lower density of professional and managerial women – most boroughs score below 10% and none had over 20% in this occupational group. The spatial patterning is also different. The suburban concentration almost disappears. What emerges instead is a concentration

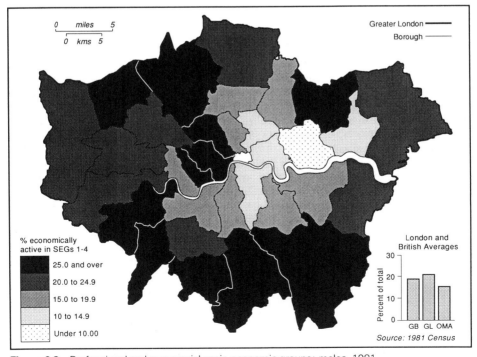

Figure 6.3 Professional and managerial socio-economic groups: males, 1981

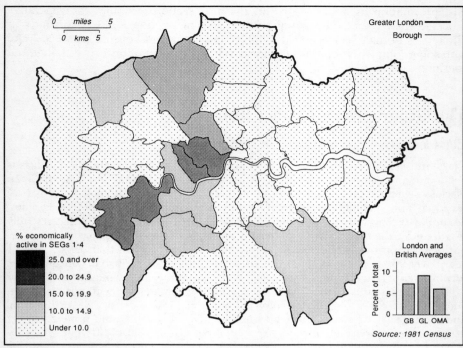

Figure 6.4   Professional and managerial socio-economic groups: females, 1981

of high female social status in west-inner London and in adjoining south-west outer boroughs.[1]

These two gender specific maps show a sort of spatial conflation with Figure 6.2 – the map of full-time female domestic workers. Where fewer women work as housewives, there is a tendency for more economically active women to be professionals or managers (although south-west suburban London – Richmond and Kingston, have relatively high levels of both). Conversely, where there are many full-time women domestic workers there tends to be many male professionals and managers (again west-inner London has fewer housewives but relatively high levels of high status males). These spatial patterns and associations imply that, in order to achieve high social status, women need to move away from that sort of situation where they are expected to be primarily domestic workers; equally, in achieving high status it is useful for men to be supported by female domestic workers. This is confirmed by gender aware studies of job mobility and household dynamics (see Morris, 1989).

The distribution of full-time domestic workers (Figure 6.2) is less class related than these maps of professional and managerial workers. But the more 'gender-specific' patterns shown in Figure 6.2 have some familiar counterparts. These are the patterns of household composition and activities usually referred to as 'demography' (as opposed to 'social structure', 'economic characteristics' etc.). Figure 6.5 gives one example, portraying the spatial distribution of 'conventional households' in London – those consisting of or centred around a married or cohabiting couple.

The inner–outer London, west–east London contrasts are marked on this map, replicating to some extent the pattern for full-time domestic workers. This index

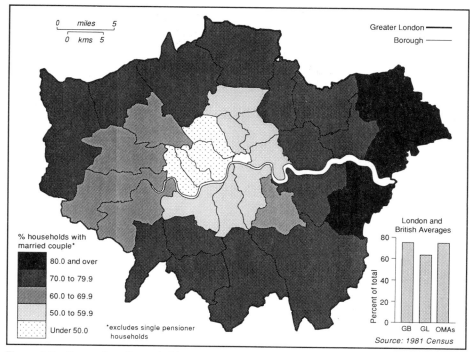

Figure 6.5 Conventional households as a percentage of total households, 1981

is also more inferential and less direct than those used up to now – some married couples may well be 'unconventional' in terms of gender relations, while some single person or non-couple households may be waiting for conventionality to happen (in census terms, cohabiting male–female couples are defined as married). Many single households are also, of course, involuntary through bereavement or divorce (although single pensioner households have been excluded from Figure 6.5). Nonetheless, it is remarkable how large the spatial variations are, again pointing to west-inner London as a site for less conventional gender relations (according to the British Social Attitudes Survey, the conventional ideal is a housewife – maybe with part-time employment when children go to school – with a husband in, or seeking, full-time paid work; Ashford, 1987). So while below 40% of households were 'conventional' in these terms in Kensington and Chelsea, over 80% were in Havering and Bexley. Note that this map of conventionality is not just a statistical mirror of our base map of full-time domestic workers ('housewives'). Fully 28% of the latter are single, widowed or divorced women. Most of these will be caring for children or other dependants – nearly all of London's 158 000 lone parent families were headed by women for instance. Similarly 30% of all London households were headed by a woman, even though the census almost invariably follows patriarchal practice in defining a woman as household head only if there is no man to fill this role.

This 'demographic' picture is mirrored by similar indices. Figure 6.6 shows, for instance, a concentration of single women below 50 (where the incidence of widowhood is low) in the same areas of west-inner London (57% and 55% of women are between 16 and 50 in Kensington and Chelsea, and Westminster, respectively, compared to just 27% in Bexley). More indicative of variation in

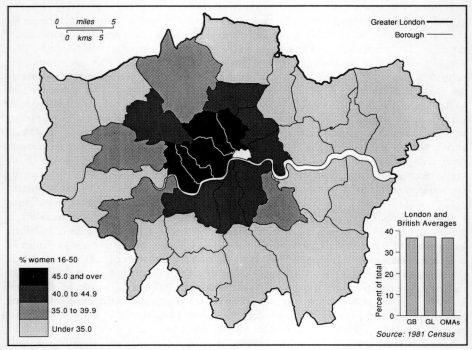

Figure 6.6   Single women 16–50 years, 1981

gender roles is Figure 6.7 which maps the proportion of women in paid work who lived by themselves. Although some of these women may have wished to become housewives, and some women in multi-person households have been economically independent, this variable is a useful index of women's relative independence. The women concerned were economically independent and were not party to a formal or informal marriage contract binding them to a male 'head of household' (marriage, or the pseudo-marriage of cohabitation, is the major institutional means by which women take on the domestic worker role). In Britain as a whole these single, economically independent women accounted for only 8.2% of households in 1981. In west-inner London rates reached over 20% (with a high of 24% in Kensington and Chelsea). Very few of these women lived in boroughs where there were high rates of full-time domestic workers (rates were only 7% in Barking and Bexley for instance). Again, south-west suburban London was a partial exception with Richmond especially scoring relatively high on both indices – although this may partly be accounted for by inner–outer variations within the borough.

It must be noted that these 'demographic' variables are not just gendered, they also relate to class divisions. Inter-borough differences in mothers in full-time paid work with pre-school children show this quite well (Figure 6.8). As we might expect by now, west-inner London scores relatively highly on this index (16% of mothers with pre-school children in high status Westminster have full-time paid jobs). However, some boroughs with high rates of full-time domestic workers also score relatively highly on this index (e.g. Newham (12%) in inner-east London). We might surmise that in areas like this, although the ascribed role for women is usually a domestic one, financial pressures push women into

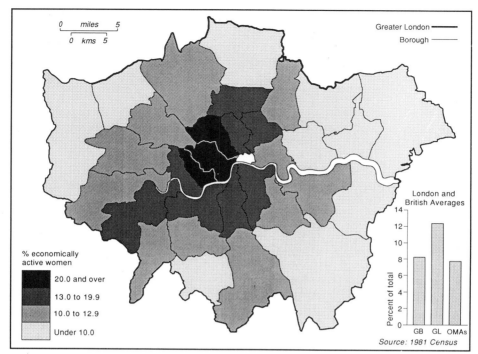

Figure 6.7   Independent single women households, 1981

the labour market – where relatively more men are unemployed or in low wage jobs. This pressure does not, apparently, extend to the suburban east of Barking or Havering. Few men living in these boroughs are employed in professional or managerial jobs, but many are in quite well-paid skilled manual jobs, while unemployment rates are much lower than in the traditional, inner East End. Hence, pressures on women, from men, to get a paid job will be less. Indeed, they may even be in the reverse direction, for the social status, and even the masculinity of British men has traditionally been heightened by their ability to support a full-time housewife (cf. Pahl, 1986). In contrast, women living in west-inner London are more likely to be in independent households and/or have access to higher wages or attractive professional jobs. Highest scoring of all, however, are two intermediate boroughs in west London – Brent and Ealing with 19% and 17% of mothers employed full-time. These two boroughs probably combine elements of both situations – relative independence for some women and household pressures to work for financial reasons. There is also likely to be an exacerbating factor which I almost completely ignore in this paper – that of ethnicity. Most black households are poorer than average, with high male unemployment, and a high proportion of Afro-Caribbean households are headed by single women. Ealing and Brent are the boroughs with the highest proportions of black residents in London.

Patriarchy and geography

How should we explain the 'demographic' characteristics illustrated in these figures? Fundamentally, they are concerned with gender relations, that is relations

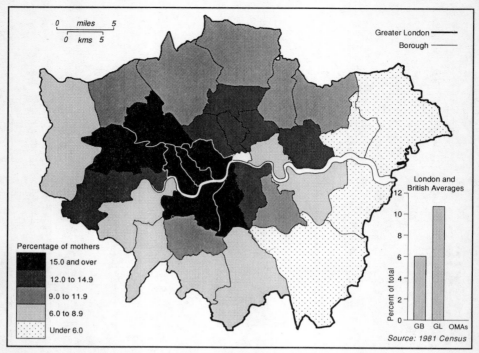

Figure 6.8   Full-time employment of mothers with pre-school children, 1981

between men and women about who does what, for whom, and what is expected in return. But how are these relations structured? This is where the concept of patriarchy becomes useful.

Patriarchy has been defined as '. . . a system of social structures, and practices, in which men dominate, oppress and exploit women' (Walby, 1989, 214). What is this domination about? It is not just a matter of gender discrimination, rather it concerns fundamentals of controlling and allocating social resources. Partly – and this is what I concentrate on in this paper – patriarchal domination seeks to control the labour of women so as to benefit men. There is a division of labour which allocates work on the basis of gender, both in homes and in paid workplaces, and which distributes rewards and status for this work. These allocations are deeply unequal with women doing more work for less reward. We should not forget that patriarchal domination is also concerned to control women in themselves, both in order to establish power over female sexuality and also to defend and define gender-based roles. Patriarchy as a social system underlies both divisions of labour and household dynamics, and so helps determine the economic, social and demographic characteristics of populations. This is why census information gives some information about gender relations and hence enables us to map proxy variables for London.

But how exactly does patriarchal domination work? Walby (1987, 1989) has identified six social dimensions which together form a self-reinforcing system of domination. These dimensions are:

1   patriarchal divisions of labour in the household;
2   patriarchal divisions of labour at paid work;

3   patriarchal practices in the state;
4   male violence;
5   sexuality;
6   patriarchal practices in civil society (e.g. in religion, the media, education, etc).

For Walby, it is the first two dimensions concerning the division of labour which are fundamental (although some feminists place more emphasis on dimensions like sexuality; see Walby, 1989). In the household women work to reproduce the labour of men, but have little control over what has been created. Men exchange their labour in the capitalist market for a wage but women do not typically receive a proportion of that wage equivalent to the value or the time of their work; still less do they control its allocation. Typically, women also receive a lesser share in the consumption of household goods ranging from food to leisure time. Partly, this results from the fact that patriarchal power is embodied in the institution of marriage (regardless of whether, or how, individual husbands choose to exercise that power). But partly this is where the interaction between household and workplace divisions of labour is crucial. For many women will have little choice; to gain access to resources they must marry or live with a man. Patriarchal practices in the workplace deny them secure, high status or high-income jobs.

This interaction demands an 'historical compromise' between the interests of patriarchs and capitalists (usefully, the latter are normally men). Capitalists are denied access to women's labour – which initially at least is usually cheaper – and a balance is set up between capitalist and patriarchal interests. This is the core of Walby's (and others') 'dual system' approach, although other approaches would prioritize patriarchy, or – indeed – capitalism (see the debate in Crompton and Mann, 1986). This compromise varies historically and regionally as different occupations are developed in different industrial sectors, which explains the sex typing of various occupations and hence regional differences in women's paid work. Finally, other dimensions of patriarchy support this settlement – assertive women may be physically attacked, state policies are often directed towards maintaining patriarchal families, girls are socialized and educated towards 'women's work', monogamous heterosexuality in husband–wife households is maintained as the norm (with convenient double standards for men), and so on.

Given that patriarchal gender relations are so important in determining social, economic and demographic characteristics, they are important in the creation of geographies (like regional variations in women's paid employment). But spatial variation in itself can make a difference to how social processes work, including patriarchy. Thus patriarchal relations operate in various contexts which affect *how* such relations work. So in industrial Lancashire, for instance, women have traditionally been employed full-time in textiles, (only in central London do fewer women work as full-time housewives; Duncan, forthcoming), where they have influenced more the 'public' spheres of trade unionism and politics conventionally reserved for men. It is in Lancashire, too, that the custom of the wife taking control of the household wage, and allocating pocket-money to the husband, was most widespread (Zweig, 1952). In industrial Lancashire the crucial and self-reinforcing interaction between patriarchal household and patriarchal workplace relations was not as strong as elsewhere (although it varied subtly within Lancashire, see Mark Lawson *et al.*, 1987). In some other parts of Britain a lack of full-time paid work for women or a particularly strong culture of masculinity reinforces gender divisions of labour. Classically in areas of coal-mining and heavy industry many more women work as full-time housewives (even in large urban areas like Newcastle). In London, the industrial 'towns' of Barking and Dagenham are more like these areas as far as gender divisions of labour are concerned. In other countries altogether different contextual variations produce different results. In non-Islamic West Africa, for instance, the penetration of capitalist modernization has increased patriarchal power, which was relatively weak in pre-existing societies. Conversely, in Islamic societies modernization can increase women's paid employment in high status jobs, while in Latin America modernization encourages their employment in low status jobs (Scott, 1986).

However, social systems do not only operate in different contexts; they themselves have geographies. They have boundaries and are themselves unevenly developed, so the strength or pervasiveness of patriarchal systems is not the same in different places (see Duncan and Savage, 1989). This is not just a matter of contingent contexts affecting how patriarchal relations work (for example, there is lower male unemployment and fewer women's jobs in Barking than in Tower Hamlets, hence contingently women's rates of full-time domestic work are higher in the former). Rather, patriarchal relations may work differently. This raises the intriguing question of when quantitative differences become qualitative differences. In west-central London, for example, where around half of working women are in full-time paid work, only about a third are full-time housewives, less than half of households centre around a married couple, and 20% or so of employed women live in single-person households – does all this mean that in this area patriarchy works differently as far as gender divisions of labour are concerned?

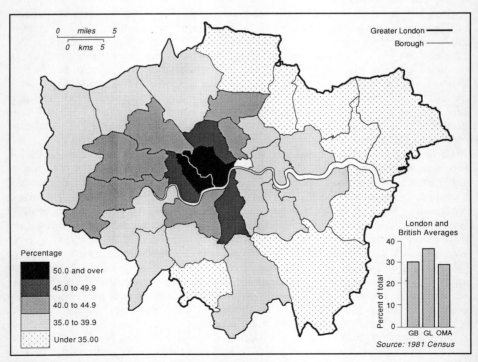

Figure 6.9   Working women in full-time paid work, 1981

Figure 6.9 maps the distribution of working women in full-time paid work in London. The usual pattern is replicated, with high values in west-central London, and low values in both high and low status suburbs, especially in the east and south. Figure 6.10 confirms that west-central London is also a high wage area for women, although interesting variations exist between lower wage suburbs and intermediate areas (London as a whole is a high wage area for women). Figure 6.11, comparing the chance of women and men being in full-time paid work, is perhaps more informative as far as gender *relations* are concerned. Women in paid work in west-central London were only 40% less likely than men to be employed full-time in 1981, but in the outer suburbs they were 60% less likely (for Britain as a whole the figure was nearly 70%).

However, these expectations of increasing gender equality in west-central London are dashed by Figure 6.12, showing women's wages as a proportion of male wages for full-time work in the 1980s. This figure and Figure 6.10, using the same sources, have to be read with caution, for the data are for place of work, not place of residence (as for the other maps), and some borough samples are small. However, the overall message is clear. There is little variation between boroughs in women's wages as a percentage of men's (the range is from 62% in Brent to 73% in Kensington and Chelsea). London, including west-central London, is little different from the national picture where women in full-time work in 1981 on average earn only 67% of the average male full-time wage. In other words comparatively many women in west-central London are in full-time paid work, and women with jobs in that area earn comparatively more than other women, but men also earn more in these areas so gender differentials remain more or less the same (and I exclude part-time work and domestic work – nearly all of which is per-

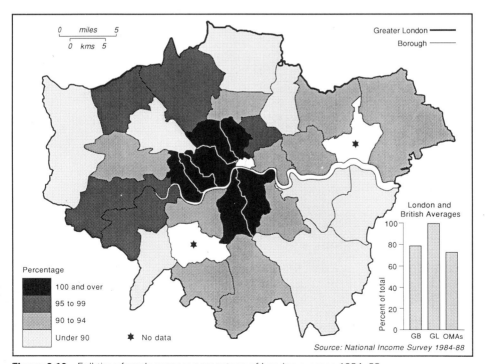

Figure 6.10   Full-time female wages: percentage of London average 1984–88

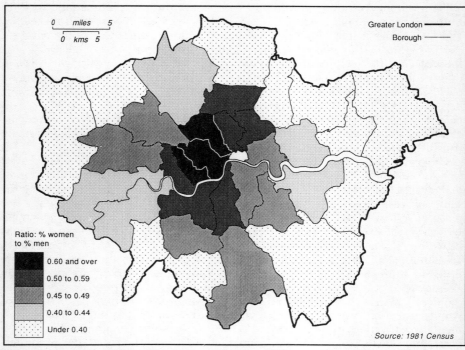

Figure 6.11   Full-time paid work: women versus men, 1981

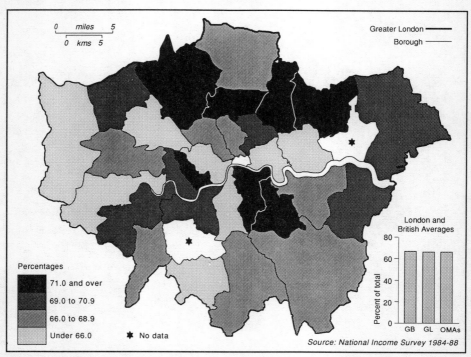

Figure 6.12   Female wages as a percentage of male wages: full-time work 1984–88

formed by women). I do not have comparative information on other dimensions of patriarchy, but we know that levels of male violence towards women are as high in west-central London as elsewhere (Kelly, 1988; HMSO, 1989). Women may do better in this area as far as access to full-time work is concerned, and the traditional male breadwinner/female homemaker household may be less common there (see Figures 6.5 to 6.7) but unequal gender divisions of labour are still dominant and it may well be that other patriarchal dimensions of control might compensate for any weakening on the labour front.

This discussion of how patriarchy may operate differently exposes the inadequacies of the ecological approach used so far. Spatially aggregated data may be heuristically useful, and a valuable starting point, but it is limited when it comes to discovering how social processes work. In addition, borough level data are spatially crude, since large variations exist within some boroughs (e.g. Brent, Croydon, Richmond) while others are distinctive, partly because they are relatively small (e.g. Westminster, Barking).

In the next section I will attempt to compensate for these limitations by examining information about processes of gender differentiation. Further research on how gender relations vary, and why, requires intensive, in-depth research in households, workplaces and civil institutions. Here an ecological approach is used to provide a spatial framework for analyzing and presenting information on processes of gender differentiation.

Constructing a spatial framework

Figure 6.13 presents graphically the relation between two key indicators of the gender division of labour. These indicators are the proportion of working women

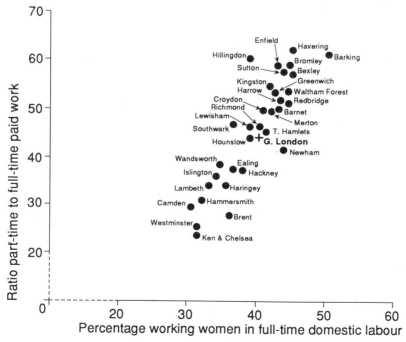

Source: 1981 Census

Figure 6.13 Gender divisions of labour: women in London boroughs, 1981

in full-time domestic labour (cf. Figure 6.2), and the *ratio* of part-time to full-time paid employment. As we might expect, the higher the proportion of housewives in any borough, the higher the percentage of part-time workers (the two indices are of course statistically independent). This relationship is remarkably regular, suggesting the logic of gender divisions which produces it, although particular boroughs show some interesting deviations. For example, Brent and Newham show a higher proportion of full-time female workers than we would expect, Hillingdon more part-time. These small deviations from the norm can probably be explained with reference to contingent, contextual factors: Hillingdon is dominated by Heathrow; in Brent there is a large Afro-Caribbean group and marriage rates are lower; in Newham more males are unemployed.

This same basic relationship emerges for Britain as a whole, although, again there are interesting deviations for particular cases (Duncan, forthcoming). Most of the London boroughs fall into a high full-time paid work/low domestic labour category. This category is also found in the 'western crescent' of economic growth around London, in industrial Lancashire, in other areas with long-established women's jobs like Leicester, and, to some extent, in central Scotland and in regional service/university towns (e.g. Brighton, Exeter). Barking and Dagenham is an exception for London, deviating towards the British average, while west-central London stands apart with the lowest national proportions of housewives and the highest rates of female full-time workers in Britain.

Using the graph in Figure 6.13, five groups of boroughs were distinguished according to the dominant labour roles for women. First of all, a basic distinction was made between boroughs where women had, on aggregate a 'paid worker' role (high rates of full-time paid work, low rates of full-time domestic

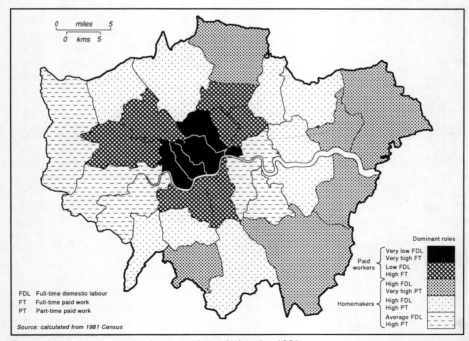

Figure 6.14   Dominant roles in women's work: London 1981

Table 6.1 London boroughs classified by dominant roles in women's work, 1981

Borough cluster	Attributes Domestic work	Paid work	Boroughs
1 *Paid workers*			
1a	very low	very high full-time	Camden, City of London, Hammersmith and Fulham, Kensington and Chelsea, Westminster
1b	low	high full-time	Brent, Ealing, Hackney, Haringey, Islington, Lambeth, Wandsworth
2 *Homemakers*			
2a	high	very high part-time	Barking, Bexley, Bromley, Enfield Havering, Sutton
2b	high	high part-time	Barnet, Croydon, Greenwich, Harrow, Kingston, Merton, Newham, Redbridge, Waltham Forest
2c	average	high part-time	Hillingdon, Hounslow, Lewisham, Richmond, Southwark, Tower Hamlets

labour) and boroughs where women had a more conventional 'homemaker' role (high rates of full-time domestic labour and part-time paid work). Sub-divisions distinguished differences in degree within these categories. These categories were determined with reference to the average figures for London as a whole. Figure 6.14 maps these categories, with the particular position of west-central London being clear. Table 6.1 gives further details of these groups.

Processes and Actions in Gender Differentiation

So far I have treated gender divisions of labour as though they consist of spatially and historically fixed aggregates. This assumption presents certain limitations in examining *how* these 1980s spatial patterns were created. The task of this section is to go further in this direction; although it will be obvious that more research is necessary to do this adequately.

Labour markets

The core of the paper so far has concerned differential male and female work. Women had much lower rates of full-time paid work in 1981, but accounted for nearly all part-time paid workers and unpaid but full-time domestic workers. These gender roles were geographically differentiated. There were significant differences between London and other parts of Britain, and within London 'paid workers' and 'homemaker' areas could be distinguished in terms of women's labour roles.

But how are these groups of workers recruited, what are the levels of mobility between them, and what relative cohesion do they have over time? Do recruitment, mobility, and cohesion also vary by gender and by area? These are classic questions in establishing the relative significance of social structure and how different groups fare within it, although up to now they have usually been applied to *class* structure. Studies have also been limited in using paid work as an index of class, and even within this 'gender-blind' parameter they have often excluded women's paid work (see Lockwood, 1986). It is important to know, for instance, if full-time domestic workers are stuck in that role or have easy access to full-time paid work. Another criticism of work on social mobility is that too often it has depended on historical snapshots, taken from sequential censuses, and so compares statistical aggregates rather than tracing the movement of individuals through social structures. The Longitudinal Study (LS), which followed a 1% sample of census respondents from 1971 to 1981, was partly designed to offset this failing and is used here to examine recruitment, mobility and cohesion in gender divisions of labour within London.

Table 6.2 Origins of 1981 full-time paid workers

| | | Status in 1971 Per cent | | | | |
		Full-time work in London	Part-time paid or full-time domestic work in London	Outside labour force in London*	Rest of England and Wales	Rest of world
Paid worker boroughs	Female	29.8	14.3	22.3	20.0	15.6
	Male	52.6	1.9	19.4	12.8	13.2
Home-maker boroughs	Female	26.9	32.1	24.0	10.3	6.3
	Male	57.6	1.7	19.6	9.5	11.7

Source: OPCS Longitudinal Study (Crown Copyright reserved)

*Includes education, retirement, sickness and unemployment

Table 6.2 presents information on the recruitment of those in full-time paid work in 1981, broken down by gender and separating 'paid worker boroughs' (groups 1a and 1b in Table 6.1) from 'homemaker boroughs' (groups 2a, 2b, 2c in Table 6.1). The first thing to emerge is the gender difference in recruitment. Male full-time workers have much greater stability of employment (over 50% in both areas were in the same group in 1981 as in 1971, compared to below 30% for women). Similarly, few full-time men were recruited from part-time paid work or full-time domestic work, whereas this category provided many recruits to full-time paid work for women.

None of this is surprising, given what we know about gender divisions of labour, although this formally confirms that our 1981 census maps were not an aberration. More surprising is the gender differentiated role played by the paid worker boroughs as a port-of-entry for recruitment. London, and especially inner London, has always been important as a source of jobs for migrants and has thus played a significant role in allowing social mobility between occupations (cf. Savage and Fielding, 1989 on the service class). Not unexpectedly, therefore, in-migration is a significant source of recruitment for full-time workers. But this role is gender-differentiated. Rates of recruitment by in-migration are much higher for women full-time workers in paid worker boroughs, compared with men as a whole and women in the homemaker boroughs. Paid worker boroughs, seem to be a crucial port-of-entry for full-time paid work for women whereas male in-migration rates hardly differ between the borough groups. Conversely, the homemaker boroughs are a major source of none-migrating labour from part-time and domestic work, with recruitment rates from this source over double those in the paid worker boroughs. This, perhaps, indicates the spatial

Table 6.3 Origins of 1981 part-time paid and full-time domestic workers

	Part-time paid work or full-time domestic work in London	Full-time paid work in London	Status in 1971 Per cent Outside labour force in London	Rest of England and Wales	Rest of world
Paid worker boroughs	49.2	17.4	14.6	5.1	13.7
Homemaker boroughs	47.3	19.4	21.9	5.6	5.8

Source: OPCS Longitudinal Study (Crown Copyright reserved)

concentration of the classic 'latent reserve army' of female workers wholly or predominantly in domestic work (Humphries, 1983).

Table 6.3 reports similar recruitment figures for part-time paid and full-time unpaid domestic workers, although this time for women only (men form an insignificant part of this category). For the sake of clarity I have lumped together part-time paid work (defined as less than 30 hours per week) and full-time, unpaid domestic work. In Britain, it is only when the female partner takes on full-time paid work that conventional gender roles become less severe (Witherspoon, 1985; Laite and Halfpenny, 1987). The relative stability of female participation in these conventional roles is more apparent in Table 6.3, as 1971–81 continuation rates of nearly 50% almost match male rates for full-time paid work (Table 6.2). This aptly reveals the homemaker/breadwinner gender distinction. There are also two significant differences between the borough groups in Table 6.3. First, paid worker boroughs recruit more full-time housewives and part-time paid workers from overseas. This probably reflects the inward movement of pre-existing dual adult households from abroad while most migrants from the rest of England and Wales will be single and enter the full-time job market (Table 6.2). Secondly, homemaker boroughs have higher rates of internal recruitment. If there is a female reserve army of labour in these areas it is largely produced *in situ*.

Table 6.2 and Table 6.3 measure what social categories in 1971 members of a 1981 category came from (i.e. recruitment). Table 6.4 and Table 6.5 use the same data in reverse, that is what 1981 social categories members of a 1971 category went *to* (viz. social mobility). I have simplified the tables in this case by excluding destinations outside London, movement out of the labour force (to education, retirement, unemployment or permanent sickness) and deaths. Conversely, I have been able to include information on migration between the paid worker and homemaker borough groups.

Table 6.4 presents this mobility information for people who were full-time paid workers in 1971. Again, differential gender mobility rates are prominent. Once men become full-time paid workers they stay as such, until they leave the labour market. The only difference between the borough clusters is that over 20% of the males in paid worker boroughs migrated to homemaker boroughs. For women, however, wastage is far higher. Indeed, around half of women full-time workers in 1971 had part-time or domestic work by 1981. Yet there is significant variation between the borough groups (nearly 60% of women in paid worker boroughs stayed in full-time paid work, but only 49% did in homemaker boroughs).

Table 6.4   Destinations of 1971 full-time workers

| | | Status in 1981 Per cent Full-time paid work | | Part-time paid work or full-time domestic work | |
		same area*	other area*	same area*	other area*
Paid worker boroughs	Female	47.4	10.9	28.9	12.8
	Male	74.6	21.3	3.7	0.5
Homemaker boroughs	Female	46.6	2.6	53.3	2.5
	Male	94.2	3.0	2.1	0.7

Source: OPCS Longitudinal Study (Crown Copyright reserved)

**area* refers either to paid worker boroughs or to homemaker boroughs

Table 6.5   Destinations of 1971 part-time paid and full-time domestic workers

| | Status in 1981 Per cent Full-time paid work | | Part-time paid work or full-time domestic work | |
	same area*	other areas*	same area*	other areas*
Paid worker boroughs	15.2	5.9	68.6	10.3
Homemaker boroughs	25.3	0.7	72.4	1.6

Source: OPCS Longitudinal Study (Crown Copyright reserved)

*as Figure 6.4

In addition, there is considerable migration to the homemaker boroughs, either as full-time paid workers (10.9%) or as part-time or full-time domestic workers (12.8%), with virtually no movement in the other direction.

Table 6.5 repeats this exercise for part-time paid and full-time unpaid domestic workers – again excluding men who hardly enter this part of the workforce. As with Table 6.3 on recruitment, the strongest impression is of relative stability (the female homemaker role replicating the male breadwinner role in this respect). However, there was also considerable mobility into full-time paid work. Surprisingly, this is greater for homemaker boroughs. However, breaking down this figure between part-time paid workers and full-time domestic workers shows that the former largely account for this differential; it is perhaps just because this group is relatively more important in the homemaker boroughs that accounts for its relatively greater mobility – when most women in the paid worker cluster with formal jobs already work full-time. Nonetheless, the paid worker boroughs again show their port-of-entry/labour export role with very little reverse migration.

Briefly, then, the Longitudinal Study (LS) data on individual work paths between 1971–81 has established three things. First, it has confirmed the overall male breadwinner/female homemaker categories in terms of recruitment and mobility. Secondly, it has illustrated the importance of the paid worker cluster as a port-of-entry/labour export area for women. Finally, it confirms that conventional gender divisions of labour, although still dominant, are weaker in the paid worker boroughs. Remember also that this dual division into homemaker/paid worker categories is crude. The paid worker category in fact lumps together a core zone (1a in Table 6.1) and a larger, more transitional area (1b in Table 6.1). We might expect these differences to be even more marked in the core zone. Similarly, the homemaker category conflates an extreme area (2a in Table 6.1) with more transitional areas (2b and 2c in Table 6.1).

Getting or leaving a full-time paid job, becoming a part-time employee or a full-time housewife are major work strategies – or perhaps responses – available to women living in dual capitalist and patriarchal societies. Homeworking may also be an option. In this dual system, these actions are as much or more a response to household negotiation as to labour market structures, and reflect major household events like marriage, childbirth, entry of children into school, the children's eventual maturity, loss of partner and so on. We should not think that these decisions are made in some economically 'rational' way, even from the viewpoint of a dominant husband. Beliefs and attitudes about appropriate work roles and behaviour are heavily gendered, leading to what can seem economically 'irrational' behaviour (e.g. women in part-time work often do *more* housework than full-time housewives, see Laite and Halfpenny, 1987). It is this interaction of beliefs, opportunities and pressures – in both households and formal workplaces – that are reflected in the patterns found in the longitudinal data. My general conclusion, therefore, is that this interaction – while of the same overall type – is different in degree in paid worker boroughs compared to homemaker boroughs. This supports the conclusions reached from the analysis of the ecological information.

Housing markets

The Longitudinal Study (LS) data allowed us to examine more closely, if still on an aggregate level, strategies for gaining income and carrying out household work. Another important household activity is to secure housing. How housing is acquired is also a gendered activity, depending on definitions of economic competence and social status. For instance, the usually low economic position of single-parents (who are overwhelmingly women), combined with judgements of high – if somewhat disreputable – social need, usually secures local authority housing, even if normally in the less attractive hard-to-let estates. Owner occupation, in contrast, usually demands a higher economic position, combined with judgements about social status (e.g. employment prospects). It is also the chief means of gaining access to housing in Britain, even for the apparently economically irrational cases of young and single mobile workers. The access of women to owner-occupation therefore gives us another clue as to their relative status over time and geographically. Do women gain access in their own right – socially defined as socially independent and economically competent persons – or via a male head of household who better incorporates these qualities?

For the years 1978, 1987 and 1988 data were collected on borrowers from the Nationwide Anglia Building Society for each of the five borough groups identified in Table 6.1. That is, the paid worker cluster used with the LS data is broken down into extreme (1a) and transitional (1b) areas, while the homemaker cluster is disaggregated into extreme (2a), transitional (2b) and residual mixed (2c) areas (see Figure 6.14). The data are taken from the records of one building society only, but the Nationwide Anglia is the second largest in Britain and currently makes about 10 000 dwelling purchase loans in London per year. I take it to be reasonably representative.

Figure 6.15 shows the overall proportion of women borrowers over time. This

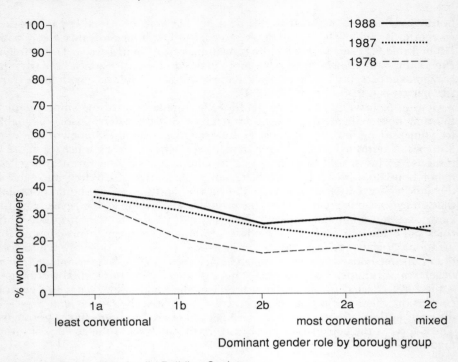

Source: Nationwide Anglia Building Society

Figure 6.15   Building society borrowers by sex and ecological area: London 1978–88

is defined, by the Nationwide Anglia, as those loans borrowed either by single women or joint mortgages where a woman's name came first (there will probably be many joint mortgages where a man's name came first but where a woman co-borrower contributes a substantial part of repayment costs – in such cases the woman is defined as secondary and it is this socially defined role we are interested in here). Two major features emerge from Figure 6.15. First, the proportion of women borrowers is low but is increasing over time, from 17.2% in London as a whole in 1978 to 28.2% in 1988. Secondly, this proportion varies considerably by borough group being consistently high in the least conventional group and conversely low in the most conventional and residual areas. Interestingly, dwelling purchase by women has been more prevalent in the least conventional borough group (ranging from 34% to 38%). But it is in transitional and more conventional areas that gains have been more marked.

We can break this information down by borrower type, distinguishing between single and married, male and female (heterosexual cohabitees are defined as married). Figure 6.16, for 1978, shows what we might call a traditional pattern. Borrowing by married men predominated in all areas, but especially in more conventional borough groups where they accounted for over 70% of loans. In less conventional boroughs, this dominance was reduced as single borrowers were more common (with 57% of loans in type 1a boroughs). This links in with the Longitudinal Survey, where paid worker boroughs (1a and 1b) were a port-of-entry for migrants. The preponderance of single loans in the least conventional areas can be seen as a life cycle effect. Note, however, that this effect was

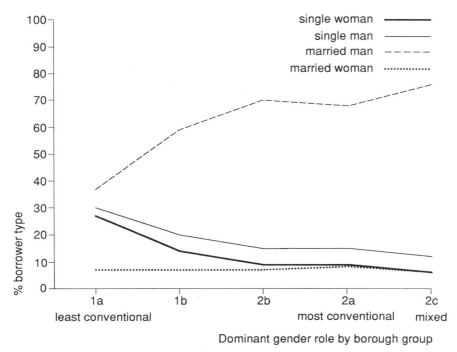

Source: *Nationwide Anglia Building Society*

Figure 6.16   Building society borrowers by type and ecological area: London 1978

gendered, with single men always outstripping single women (particularly in the most conventional boroughs; Figure 6.16). Indeed, in the most conventional areas borrowing rates for single women were at the very low level recorded everywhere by married women.

Figure 6.17, presenting the same information for 1987, shows considerable changes. Married men still predominated in the most conventional areas, although with reduced rates of 40–50% of borrowers. Married women still took up few house loans. But borrowing by single people had increased everywhere, attaining overall majorities in borough types 1a, 1b and 2c, the least conventional and mixed areas. However, this increase was strongly gendered. Only in the least conventional boroughs (1a) did single women approach rates for single men (30% and 39%, respectively). A contemporary survey of flat conversion in the paid worker boroughs of Camden, Hackney and Haringey gives a similar impression, for women were named as the primary convertor in 45% of cases (Barlow, 1988). Elsewhere rates for single women were considerably lower than for single men. Indeed, while the life cycle effect is disappearing for single men (with considerable increases in borrowing rates in more conventional areas), it remains strong for women.

In terms of building society lending, 1988 was an abnormal year (Figure 6.18). An unprecedented price boom, and then budget changes which took effect in August, made it financially more difficult for single people to buy dwellings. Combined with a continuing reduction in renting opportunities, buying of almost panic proportions by single people resulted. Not surprisingly, there-

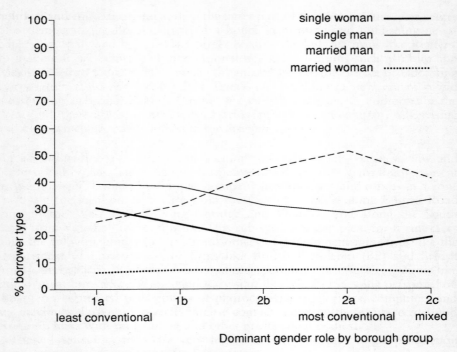

Source: *Nationwide Anglia Building Society*

Figure 6.17   Building society borrowers by type and ecological area: London 1987

fore, borrowers in all borough groups were dominated by single persons. Even so, basic gendered patterns did not change. Married men were most important in the most conventional boroughs, married women everywhere made up few borrowers, and the life cycle effect was most marked in the least conventional borough groups. There is one important, gender-specific, qualification, however. Only single women still showed this life cycle effect, with high borrowing in the least conventional areas, and lower rates in the most conventional. For single men *no* life cycle effect remains, with rates of borrowing more or less constant in all areas. It was single men who replaced married men in the abnormal conditions of 1988.

This building society borrowing data suggests three main conclusions. First, the difference between married men and married women is testimony to the head of household/breadwinner role achieved by men on marriage. Secondly, there is a life cycle effect with single people more commonly borrowers in the least conventional areas, married men more in the most conventional. However, thirdly, this life cycle effect is strongly gendered with single men predominating over single women even in less conventional areas. Indeed, in the abnormal conditions of 1988, already apparent in 1987, this life cycle effect disappeared completely for single men. In contrast the least conventional areas remained the most important sites of house purchase by single women.

These conclusions fit in well with both the LS data and the ecological data. The LS data showed the paid worker boroughs to be particularly important for women as a port-of-entry. The ecological data showed that in these boroughs women achieved greater access to full-time, relatively well-paid employment, and

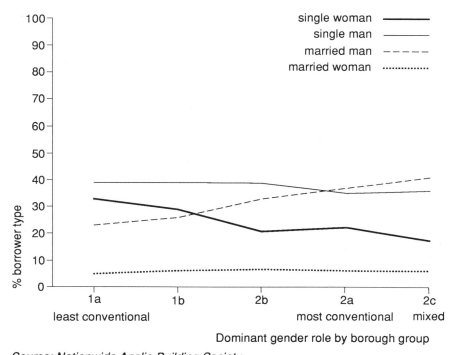

Source: *Nationwide Anglia Building Society*

Figure 6.18 Building society borrowers by type and ecological area, London 1988

that remaining outside conventional, married households was often bound up with this superior access. When economic conditions changed in 1987–88, single people replaced married men in dwelling purchase. But this life cycle change was heavily gendered; in all but the least conventional gender role areas it was single *men* who replaced married men. Short-term economic change would have had less effect on patriarchal gender relations. Borrowing by married women remained low everywhere, and borrowing by single women remained concentrated in areas of less severe gender divisions of labour.

Perspective

The LS data on socio-geographic mobility and the information on dwelling purchase has allowed us to examine more closely *how* the geography of gender roles has been created. The recruitment and cohesion of different groups of workers is strongly gendered, and this gendering is spatially differentiated. The same goes for access to housing, at least in the dominant owner-occupied sector. Labour market and housing market structures (with unpaid domestic labour in the former) are two basic contexts for social behaviour. There are, however, other important structural contexts – for instance those surrounding sexuality – which have not been considered. Neither women nor men are passive agents in these socio-geographical structures. Both can negotiate, resist or compel changes, using both ideological and material resources; although men commonly have more resources to draw upon. One significant change in this equation, however, is that provided by 'second-wave feminism', allowing many women to re-assess their roles and relationships to men (Coote and Campbell, 1982). This paper does not give direct information on this type of behavioural change. All we can do here is signal the need for research into how these changes also create geographies of gender.

Conclusion

This chapter has demonstrated two major points. First, gender divisions of labour are both marked and relatively fixed. Secondly, these divisions are differentially developed in different areas. In some extreme areas, such as west-central London, new gender roles may be developing in contrast to the established male bread-winner/female homemaker roles. As such, these gender divisions of labour form an essential part of the geography of London, both in explaining its current character and also in understanding how this will change in the future.

There are major limitations to the work presented here. First and foremost, the analysis remains at an aggregate, statistical level, although I have tried to include process information to enrich the ecological analysis. This might be excused as a first stab at the problem, but weaknesses in scope and depth remain. In terms of scope, this sort of information is almost entirely focused around workplace labour categories and households as units. The paper says little about internal household and workplace relationships, still less about male violence, sexuality, ideology, state behaviour and other elements of patriarchal social systems. Partly as a consequence, in terms of depth, the chapter focuses on gender *roles* but says little about the gender *relations* that produce them. It may be possible to discern that gender roles in west-central London are different to those elsewhere, but we can say little on whether gender relations are more or less patriarchal as a consequence. Finally, the paper proceeds with a good dose of historical amnesia. Its world begins with 1971 and ends in 1988. But gender divisions have not been created overnight and the historical process of creation can give many clues to their understanding.

There are perhaps three ways in which research like this could be improved. First, even aggregate and statistical information is usually interpreted in a gender-blind manner. Yet much is collected in a way that relevant information can be dis-interred from the raw data (voting surveys are one case in point). Secondly, social survey and ethnographic investigation can reveal much about internal workplace and household gender relations, as many studies show (Morris, 1989; Purcell 1989). These studies can be extended into less sociologically conventional areas like male violence and sexuality. Thirdly, there is a mass of historical material which is waiting to be re-interpreted – both original statistical material as well as survey results (cf. Pahl, 1986). Granted, this work has to be disengaged from the discourses of the time, but in this historical work alone there is at least enough for another paper on the development of gender roles and relations in London. Improved statistical information and results from social and ethnographic surveys should produce two or three more.

Appendix – Census Information and Gender Relations

The UK censuses and gender relations

The UK national census has not been designed to record gender relations. However, information on household composition and labour market position allows inferences to be made, notably about gender divisions of labour. Even here important aspects like internal household divisions of labour remain unrecorded in any direct way, while other dimensions of gender relations, like male violence or sexuality, remain entirely unrecorded. Patriarchal assumptions in both the design and gathering of the census also lessen its value. For instance, forms used by 1981 census-gatherers specified a category of 'housewife' with no corresponding male category. It is likely that many casually or unemployed women were recorded as housewives while men in similar positions were recorded as belonging to the labour market. These limitations of scope, depth and accuracy should be remembered in assessing our results.

Spatial scale

Note the usual problems introduced by spatial scale. Our maps are presented for 32 boroughs, but there is variation within these units as recorded by the 755 wards or the more numerous enumeration districts, while the boroughs are of different sizes and populations. Finally, of course, the presentation of spatially aggregate data cannot predict what any particular individual or group within that area might be like.

Defining full-time domestic workers

A special note is necessary for the definition of full-time domestic workers — a category which does not appear in the census as such. The published census information categorises the adult population under retirement age (16–65 years old for men, 16–60 for women) as 'economically active' or 'economically inactive'. The former includes not only those in paid employment, but also unemployed persons, those seeking work or waiting to start a job. The economically inactive include full-time students, retired people in this age bracket, permanently sick and 'other inactive'. This category 'other economically inactive' is made up of two categories distinguished on census forms but lumped together in the published accounts — 'housewives' and 'persons of independent means' who were not employees/employers, unemployed or permanently sick (see OPCS, 1981). However, given that there are only 14 000 men in London in this 'other economically inactive' category, compared with 928 000 women (and many fewer women than men below retirement age have independent means) I have concluded that for females this category is effectively the same as 'housewives' (i.e. full-time domestic workers). It is likely, however, that some of these women may have casual paid employment, although for many it would be accurate to define their social role as that of housewife. Similarly, it is possible that some househusbands exist.

Acknowledgements

First, thanks must go to my co-researcher Susan Halford, who gave me the idea of measuring spatial variations in gender roles and who has always been ready with help, advice and encouragement. Thanks also to Mike Savage and Andrew Sayer for comments, and to Mark Baigent, Kevin Fielding, Tony Fielding, Richard Johnson and Sue Justice for their help in collecting statistical material, to the LSE Drawing Office for their cartographic work and to the ESRC (Project No. R00231024) and the LSE for financial support.

Notes

1   Only 4.3% of this occupational group work part-time, although most of those who do will be women.
2   I give this issue greater (but still inadequate) attention in a larger version of this chapter (Duncan, 1990).

7

The Emerging Retail Structure
Barrie S. Morgan

Introduction

The retail sector of the British economy, as elsewhere in the developed world, has changed out of all recognition in the last quarter of a century. Chief among these changes, which have been fully documented elsewhere (for example, Howard and Davies, 1988), are the increased concentration of ownership attendant on the expansion of multiple outlet firms at the expense of the independent retailer, mergers and acquisitions in the corporate sector, the development of new types of shop and the emergence of new retailing formats. The total number of shops declined by nearly one-third between 1966 and 1986. Those operated by national or local multiples comprised 41% of the total in 1986 compared with 19% some 20 years earlier. Furthermore, the size of multiple outlet firms has been increasing. By 1984, there were 55 businesses with 200–499 outlets, and 21 with over 500; by 1987 there were 30 businesses with over 500 outlets, including 16 with over 1000 (Howard and Davies, 1988). New, specialized stores have become commonplace as old-established multiples have increasingly targeted specialist retail sectors and new companies have grown by successfully identifying niches in the market.

These structural shifts, particularly in the last decade, have coincided with, and in part been driven by, a rapid growth in retail expenditure and a demand for new products (such as video cameras and recorders, home computers, and freezers), which have been developed by technological innovation. The growth in home ownership and in household formation has been influential in the rapid growth of do-it-yourself stores. Consequently, consumer expenditure, which has increased by more than 50% since 1970, has grown faster in some sectors than others: whereas spending on food has been relatively inelastic. However, the market has not expanded and changed in a geographically uniform manner: in addition to regional variations, population, number of households, and prosperity have grown in metropolitan rings to the detriment of urban cores.

Changes of a similar type and scale in North America have had a profound effect on the spatial organization of retailing, notably with the development of out-of-town regional shopping centres, which have added a new tier to the intra-metropolitan retail hierarchy (Vance, 1962). That they have not had a similar impact in the United Kingdom is partly a testament to the inherent conservatism of most planners, local politicians, and, insofar as incursion into the green belt is an issue, government ministers. Nonetheless, significant changes have taken place: superstores, retail warehouses and, increasingly, retail parks are now common in the urban scene; and out-of-town regional shopping centres are making their first appearance. It might be predicted that the London metropolitan region would be in the vanguard of these developments as diffusion filters downward through the size hierarchy of cities (for example, Pederson, 1970; Robson, 1973). This, however, is not the case. The London region is a laggard in the adoption of town centre development schemes (Schiller and Lambert, 1977), superstores (Davies and Sparks, 1989), retail parks (Hillier Parker, 1989) and out-of-town regional shopping centres.

Notwithstanding the pace of development, changes in the structure and spatial organization of retailing in London largely reflect national trends. Comprehensive descriptions of retailing in the metropolitan area have been provided recently by the London Planning Advisory Committee (LPAC, 1987) and the London and South East Regional Planning Conference (SERPLAN, 1987). Rather than replicate their efforts, this chapter is selective in its approach. It focuses on three components of retail change in the London metropolitan region which are to a degree distinctive as a consequence of the size and wealth of the capital's population and its political fragmentation. A section of the chapter is devoted to each. First, changes over the last 25 years in the hierarchy of retail centres in the London region are reviewed; this area containing seven of the 40 largest shopping centres in the country (Schiller and Jarrett, 1985). Secondly, the policies of 24 of the 33 local authorities in Greater London with respect to shopping centres and out-of-centre retail development are examined, as a prelude to assessing the extent to which these attitudes are influential in the distribution of superstores and retail warehouses in the city. Thirdly, the processes of out-of-town regional shopping centre development are reviewed for the London metropolitan area, a region in which 10 million people live within 30 minutes drive of an orbital motorway.

The Hierarchy of Retail Centres

This section traces the development of the hierarchy of shopping centres in the metropolitan region defined by the 30 minute off-peak drive time isochrone from M25 junctions (SERPLAN, 1987). Following SERPLAN (1987), outside central London (the City of London, the Borough of Westminster, and the Borough of Kensington and Chelsea), the region is divided into three zones (inner London, outer or suburban London, and the metropolitan fringe) and five sectors (Figure 7.1), which together form 15 areas when they are superimposed.[1] The Boroughs of Haringey and Newham are included in inner London for this chapter.

New investment in retail stock prior to the 1980s was overwhelmingly concentrated in existing shopping centres. In total some 1.4 million square metres of new town-centre floorspace was constructed in the region between 1971 and 1984. The total gross floorspace constructed in schemes (excluding superstores and retail warehouses) in excess of 3700 square metres is listed in Table 7.1. Twenty-two centres attracted developments in excess of 23 225 square metres of retail floorspace; four of these developments (in Luton, Maidstone, Basildon and Wandsworth), each with more than 50 000 square metres of new floorspace, may be considered the equivalent of in-town regional shopping centres. Additionally, an in-town regional shopping centre of 70 600 square metres was opened at Brent Cross in 1976, although it was at an out-of-centre location at the junction of the southern end of the M1 and the North Circular road. The most striking

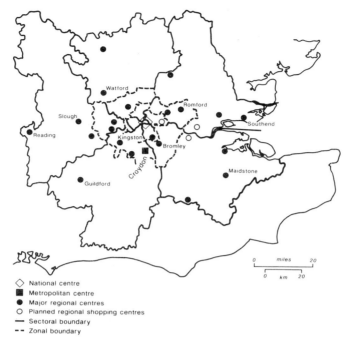

Figure 7.1   The geographical framework and the top of the retail hierarchy

feature of the zonal distribution is the relative underprovision in outer London. In 1984, this zone had 81.9 square metres of new floorspace per 1000 population, compared with 92.7 square metres per 1000 in the inner city and 150.8 square metres per 1000 in the metropolitan fringe. In general, development schemes decreased in size but increased in frequency out from inner London. For example, new floorspace was constructed in only five of the 14 centres[2] in inner London compared with 80% in the metropolitan ring, but on average the schemes in inner London were twice as large.

Given the uneven distribution of new floorspace and the decentralization of population in the region, it is to be expected that substantial changes would have taken place in the retail hierarchy in the last 20 years. Unfortunately, it is not possible to assemble a data set which enables these changes to be studied using a standard measure of rank-order. A range of indicator variables have been employed in the literature for this purpose. Number of chain stores, the presence of key stores such as Woolworth and Marks and Spencer, and the number of cinemas and banks were all used in early papers on the retail hierarchy in London (for example, Smailes and Hartley, 1961; Carruthers, 1962). Numbers of retail establishments, business types, and retail floorspace are also commonly employed indices. However, retail turnover was probably the pre-eminent measure used for British shopping centres until the cancellation of the Census of Distribution in 1981 curtailed a series dating back to 1950. This paper, therefore, employs Schiller and Jarrett's (1985) ranking by multiple branch score in 1984 as a measure of the present hierarchical position of shopping centres. The score for each centre is simply how many out of 107 selected multiple outlet firms it contained. The ranks of centres in 1961 and 1971 are based on their retail turnover.

Schiller and Jarrett rehearse the arguments for an index based on the number of multiple stores selling durable goods. Compelling as these are, there remains the danger that apparent changes in rank position are merely artefacts of measurement. This is particularly a problem in the case of the centres in central London – the West End (at the top of the regional and national hierarchy), Kensington High Street, Knightsbridge and Chelsea. These centres are distinctive since they perform a national and international, in addition to a regional, role. It was estimated, for example, that 17% of total sales in Oxford Street in 1985 came from overseas visitors; and in the 1970s when the pound sterling was weaker tourists may have accounted for almost a third of sales. Given the high volume of sales, it is no surprise that central London, and even the West End itself, comprises a separate area with distinctive characteristics and functions. Large department stores and chain stores dominate Oxford Street and Regent Street. Other areas, such as Bond Street with its haute couture and art galleries, King's Road (Chelsea) with its boutiques and antique shops, are more dominated by specialist, independent retailers. While there is no doubting the pre-eminence of the West End between 1961 and 1984, the lack of retail turnover figures for the 1980s undermines the ability to compare changes in its rank-order with those of other centres. Further symbolizing the special character of central London, retail turnover in Knightsbridge is dominated by one store (Harrods), with the consequence that the centre appears to have slipped from 3rd to equal 40th place in the regional hierarchy between 1971 and 1984. Similarly, Chelsea, where specialist independent retailers maintain a strong presence, dropped from 12th to 40th. In neither case does this decline represent a realistic picture of the changing roles of these centres. Hence, in recognition of these specialized functions, the centres in central London are excluded from the analysis of change.

The rank positions in 1984 of the top 78 shopping centres in the region outside central London are listed in Table 7.1, along with their rank positions in 1961 and 1971. Although the author is satisfied that outside central London the change in the measurement of rank-order has a limited impact[3], Table 7.1 should still be interpreted cautiously. It is not possible to rank 23 centres in 1961 and 16 in 1971 because they were too small for details of their turnover to be included in the Census of Retail Distribution; centres that were among the top 77 in 1961 and/or 1971 but not in 1984 are excluded.

Centres in inner London lost rank between 1961–71 and 1971–84, but most noticeably in the latter period when the average centre went down 17.0 places. There is a striking contrast in the period 1971–84 between inner London centres with and without new floorspace: the former lost on average only 2.4 rank-order places, with Wandsworth, Stratford and Hammersmith marginally improving their position, compared with the latter's average loss of 25.1 The worst performing inner city centre has been Brixton, which was the fourth most important centre in the region in 1961, but only equal 76th in 1984. Only one of the 14 centres in inner London, Lewisham in the south-east, was classified by Schiller and Jarrett (1985) as a major regional centre (centres with a multiple branch score of between 30 and 64). There is one minor regional centre (a score of between 16 and 29) in each sector in inner London – Peckham in the south-east, Putney in the south, Hammersmith in the west, Wood Green in the north and East Ham in the north-east. This means that, in marked contrast to the situation in the other two zones, over half of the centres in inner London aspire to only district status.

The average suburban centre also lost rank in both time periods, but the changes here were of a smaller order. Whereas the most successful shopping centres in the inner city have been struggling to hold or slightly improve their rank, the

Table 7.1 The rank order of shopping centres 1961–84 and Gross Floor Space constructed (sq m) 1970–84

South-eastern Sector

Region	Centre	1961	1971	1984	New Flrsp
Inner London	Lewisham	11	10	12	
Inner London	Peckham	13	23	42	35120
Outer London	Bromley	9	7	5	
Outer London	Woolwich	18	21	31	
Outer London	Bexleyheath	43	50	35	37160
Outer London	Orpington		57	54	13935
Outer London	Eltham	48	51	58	
Metro-politan Ring	Maidstone	19	11	9	55368
Metro-politan Ring	Chatham	29	20	11	53046
Metro-politan Ring	Tunbridge Wells	22	18	17	
Metro-politan Ring	Gravesend	28	28	42	24711
Metro-politan Ring	Dartford	46	43	50	12170
Metro-politan Ring	Sittingbourne			58	
Metro-politan Ring	Gillingham	50	55	65	
Metro-politan Ring	Tonbridge			72	8825

Southern Sector

Region	Centre	1961	1971	1984	New Flrsp
Inner London	Putney	45	46	49	
Inner London	Clapham Jnct	16	25	58	
Inner London	Tooting	31	32	58	
Inner London	Wandsworth		62	65	
Inner London	Streatham	34	44	65	
Inner London	Brixton	4	15	76	49237
Outer London	Croydon	2	1	1	4830
Outer London	Kingston	1	2	2	37067
Outer London	Sutton	14	17	17	13935
Outer London	Richmond	38	29	47	
Outer London	Wimbledon	49	45	65	
Metro-politan Ring	Guildford	15	6	4	
Metro-politan Ring	Epsom		60	28	
Metro-politan Ring	Crawley	37	34	31	
Metro-politan Ring	Woking	50	53	35	
Metro-politan Ring	Horsham			41	
Metro-politan Ring	Walton		61	76	
Metro-politan Ring	Dorking			76	

Western Sector

Region	Centre	1961	1971	1984	New Flrsp
Inner London	Hammersmith	23	49	46	
Outer London	Hounslow	21	14	14	
Outer London	Ealing Brdway	20	39	20	
Outer London	Uxbridge	44	54	30	
Outer London	West Ealing	40	36	50	
Metro-politan Ring	Reading	3	3	4	13935
Metro-politan Ring	Slough	17	16	28	22296
Metro-politan Ring	Staines	41	42	31	14864
Metro-politan Ring	Windsor			35	27034
Metro-politan Ring	Aylesbury			41	
Metro-politan Ring	High Wycombe	35	24		14864
Metro-politan Ring	Maidenhead				
Metro-politan Ring	Bracknell			76	
Metro-politan Ring	Camberley				
Metro-politan Ring	Aldershot				
Metro-politan Ring	Farnborough	42	40	76	

Northern Sector

Region	Centre	1961	1971	1984	New Flrsp
Inner London	Wood Green	12	12	28	27870
Inner London	Holloway	24	31	58	
Outer London	Brent Cross		26	12	26105
Outer London	Harrow	27	37	26	18580
Outer London	Kilburn	24		34	
Outer London	Enfield	39	48	38	
Outer London	Wembley	33	19	42	
Outer London	Golders Green	52	56	72	
Metro-politan Ring	Watford	6	4	5	30936
Metro-politan Ring	Luton	7	13	8	40040
Metro-politan Ring	Stevenage		41	20	22203
Metro-politan Ring	St Albans	30	27	23	14028
Metro-politan Ring	Hemel Hempstd	51	52	26	10962
Metro-politan Ring	Letchworth			31	8361
Metro-politan Ring	Welwyn			38	18580
Metro-politan Ring	Dunstable			38	13006
Metro-politan Ring	Hitchin			42	23225

North-eastern Sector

Region	Centre	1961	1971	1984	New Flrsp
Inner London	East Ham	26	30	50	
Inner London	Stratford		58	55	23225
Inner London	Dalston	47	47	72	
Outer London	Romford	5	5	7	37160
Outer London	Ilford	10	9	17	4738
Outer London	Walthamstow	36	38	72	
Metro-politan Ring	Southend	8	8	9	13935
Metro-politan Ring	Harlow	54	33	14	32236
Metro-politan Ring	Basildon	55	35	20	63358
Metro-politan Ring	Chelmsford	34	22	23	25083
Metro-politan Ring	Grays		59	65	15328
Metro-politan Ring	Brentwood			65	6968

Source: Schiller and Jarrett, 1985, SERPLAN, 1987

suburban zone contains centres which have performed well and badly. Five of the 24 suburban centres (Eltham, Woolwich, Tooting, Golders Green and Walthamstow) appear to be locked into long-term decline in that they lost rank in the periods 1961–71 and 1971–84 (Table 7.1). It is noteworthy that four of these are in Victorian suburbs. Elsewhere the picture is one of centres waxing and waning within a hierarchy which maintains a degree of long-run stability. Only Bromley increased its rank in both time periods – from ninth in 1961 to fifth in 1984. As was the case in the inner city, centres which benefited from new building improved their position at the expense of their neighbours: those with over 9290 square metres of new floorspace jumped on average 8.7 rank places between 1971 and 1984. This makes Bromley's success even more impressive, since it has not enjoyed the advantage of new retail floorspace (a development of 36 230 square metres will open in 1991). The most successful centres since 1971 – Uxbridge, Ealing Broadway and Bexleyheath – improved their positions by 24, 19 and 15 places respectively. However, much of this improvement, all in the case of Ealing Broadway, merely represents a recovery of ground lost to other centres in the 1960s (Table 7.1). Overall, Croydon is the most important centre in the suburbs, being the only one in the region outside the West End accorded metropolitan status (a centre with a multiple branch score between 65 and 99) by Schiller and Jarrett. This means that each sector in the suburbs has at least one major regional centre, with three sectors having two (Figure 7.1). But the southern sector clearly enjoys the best provision, as it houses major regional centres at Kingston and Sutton in addition to Croydon.

Shopping centres in the metropolitan ring have prospered since 1961. However, the changes in average rank of the zone's centres – an increase of 4.2 in the first period and 3.1 in the second – paints only a partial picture. Fifteen centres were too small for retail turnover figures to be published for them in 1971 and were assigned a rank for the first time in 1984; and it was not possible to study changes between 1961 and 1971 for another three. Over a third of the centres for which full data are available improved their rank in both periods; only two – Gillingham and Luton – lost rank. Luton's decline, from being the 8th to 15th most important centre in the region, occurred despite the addition of 81 750 square metres of new floorspace. Indeed, Luton has been unable to absorb this additional floorspace, for in 1988 15 000 square metres of retail floorspace, 13.7% of the centre's total, was vacant (SERPLAN, 1989c). There were twelve major regional centres in the metropolitan ring in 1984, with two or three in each sector apart from the southern, where Guildford stood alone (Figure 7.1). It has been noted that this sector has an over-representation of high order centres in the suburban zone, and it may be that its residents look more towards centres in Greater London, and particularly to Croydon.

The fortunes of the five sectors have also been variable. Centres in the southern, south-eastern and northern sectors have on average lost rank in both time periods, the southern sector drastically so between 1971 and 1984. The centres in the north-eastern sector fared well in the 1960s, but it is the western sector along the London–Bristol growth axis which has thrived in the 1970s and 1980s (Table 7.1). The average rank-order of its centres increased by 4.9 places, while six were ranked for the first time in 1984. Despite these trends, major regional centres were relatively evenly distributed in 1984 (Figure 7.1). The western, north-eastern and south-eastern sectors had five, the southern sector four (one of which is classified as a metropolitan centre) and the northern sector three. However, the top four centres are in the south-western arc between Croydon and Reading.

Changes in the rank-order of a shopping centre should be reflected in its retail provision. Central place theory predicts that high order centres will provide all the goods that are available in lower order centres plus a discrete range of goods of their own with a higher threshold.[4] As a centre sinks down the hierarchy so it should lose high threshold functions; as a centre rises it should gain them. In the light of this theory, it is instructive to compare the changes in shopping provision in four centres in south London attendant on shifts in ranking: Brixton and Peckham in inner London declined respectively from 15th to 76th and from 23rd to 42nd in the metropolitan hierarchy between 1971 and 1984; Orpington and Bexleyheath in the suburbs rose from 57th to 54th and 50th to 35th respectively. All but Orpington were designated as strategic centres in the Greater London Development Plan (GLDP).

These centres are compared by examining changes in the number and composition of multiple stores. The author has recently ranked the thresholds of 120 multiple stores, which for present purposes are divided into six threshold groups (Morgan, 1987). The group boundaries, in ascending order of threshold, correspond approximately with the entry into the retail hierarchy of electrical goods retailers, footwear shops, women's wear retailers, jewellers and department stores. Both the inner city centres lost multiple retailers at a time when chain stores were becoming increasingly dominant in British high streets (Table 7.2). However, while few new retailers established themselves, the withdrawal of high threshold multiples seems to have lagged behind the decline in centre importance. Consequently, 29% of Brixton's meagre 21 multiples, and 38% of Peckham's 29 outlets, are in the two highest threshold groups. This contrasts markedly with the situation in Orpington which, although it is now above Brixton in the hierarchy, has had difficulty attracting high threshold multiples: less than 10% of its 52 multiples are from the two high threshold groups, a smaller percentage than in 1970. In contrast, 35 of the 40 multiples with medium to medium/low

Table 7.2 Multiple stores in four centres by level of threshold, 1970 and 1984

Threshold		Brixton				Peckham			
		1970	L	G	1984	1970	L	G	1984
Low	1	1	1	1	1	3	1	0	2
	2	3	1	1	3	8	1	1	8
	3	3	2	6	7	5	2	2	5
	4	5	2	1	4	6	3	0	3
	5	8	4	0	4	8	3	2	7
High	6	4	2	0	2	4	1	1	4
Total		24	12	9	21	34	11	6	29
		Bexleyheath				Orpington			
Low	1	2	0	2	4	3	2	2	3
	2	10	3	0	7	8	0	11	19
	3	5	3	5	7	6	0	10	16
	4	5	3	3	7	4	1	6	9
	5	5	1	7	11	3	2	3	4
High	6	0	0	6	6	0	0	1	1
Total		27	10	25	42	24	5	33	52

Key: 1970   Number of multiple stores in 1970
      L   Number of multiple stores closing 1971–84
      G   Number of multiple stores opening 1971–84
      1984   Number of multiple stores in 1984

thresholds (groups 2 and 3 in Table 7.2) are represented in the centre. The number of multiple stores in Bexleyheath has increased at less than half the rate of Orpington, but Bexleyheath has succeeded where Orpington has failed in attracting the high threshold outlets, which now comprise 40% of the total compared with 18% in 1970. Clearly there is no simple relationship between shifts in the rank-order of centres in London and changes in their retail composition.

Planning for Retail Change

Shopping centres

The patterns of retail change outlined in the previous section cannot by any stretch of the imagination be described as planned; the best that can be said is that prior to the 1980s development was largely confined to existing centres. Nationally, Development Control Policy Note 13, which was first issued by the Department of the Environment (DOE) in 1972 and revised in 1977, emphasized the importance of maintaining the existing retail hierarchy. Retail planning decisions in most London boroughs have been taken within the context of policies laid out in Local Plans formulated within the strategic framework of the GLDP. Moreover, prior to its demise, the GLC had the right to be consulted on all developments and retained the power of direction in specific circumstances. Hence, the GLDP, which was only replaced as the strategic plan for London in July 1989, upon the publication of the Secretary of State for the Environment's Strategic Advice for London, has been a primary policy statement for retailing in the last two decades.

The GLDP, first submitted to the government in 1969 but only finally approved after substantial changes in 1976, recognized 28 centres of 'strategic importance to London as a whole' for major development and concentration of services (GLDP, 1976, 80). The centres were selected so that there was one in most boroughs. The outcome for two strategic centres, Brixton and Peckham, has been described above. Six others lost rank to non-designated centres, with three losing more than 10 rank-places. The declining fortunes of at least six of these centres (Brixton, Peckham, Clapham Junction, Holloway, Woolwich and Walthamstow) was apparent by 1971, but no attempt was made to distinguish them or to address the problem of how to reverse their decline. The criticism of the strategic centre policy by the Layfield Panel of Enquiry into the GLDP seems remarkably prescient in retrospect. 'The policies which relate to the encouragement of growth or development in them should be . . . positive, i.e. *they should not be based on* the principle that growth and development will take place in the desired centres if the planning authority stops it taking place elsewhere' (HMSO, 1973, 540, my emphasis). The proposed modifications to the GLDP, which had only reached the Deposit Stage at abolition of the GLC, recognized that its previous retail policy had been only partially successful and that local circumstances warranted a more diverse approach.

The contents of Local Plans in the 24 London boroughs which have prepared these statutory instruments have been examined to investigate local attitudes to planning town centres. It must be emphasized that the findings reported here are based on an element of subjective judgement which is inevitable when written statements of policy, with subtle variations of emphasis, are categorized. All but one, Kensington and Chelsea, are committed to planning within the framework of the existing hierarchy of centres. Six authorities explicitly commit themselves to 'preserving or maintaining the existing shopping hierarchy'; five indicate a desire for growth to be concentrated in specified centres. Tower Hamlets commits itself to developing a new strategic centre at Whitechapel, but Hillingdon is unique in its desire to promote '. . . a new shopping hierarchy . . . which reflects the needs of the late 1980s and 1990s' (Central Hillingdon Local Plan, 1987, paragraph 6.7). Unfortunately, there is nothing like the same commitment to tackling the environmental and traffic problems that are the inevitable consequence of promoting development in existing centres. Six policies with the potential to minimize this impact were examined: these were the construction of local roads to alleviate vehicular/pedestrian conflict, pedestrianization, environmental improvements, improved servicing arrangements for shops, a commitment to provide adequate off-street parking for shoppers, and a commitment to seek improved access to centres by public transport. The least commonly adopted approaches are new road-building and action to put pressure on transport authorities to improve public transport (each taken up by only 10 boroughs). Most common are environmental improvements (18 boroughs) and pedestrianization (15 boroughs). Few boroughs are committed to a comprehensive programme of improvement. Only Islington and Kingston adopt a positive stance in all six spheres; another four (Croydon, Harrow, Sutton and Merton) have positive policies in five spheres. At the other end of the spectrum, three authorities were committed to only one positive policy for improvement.

One indication of the unsatisfactory nature of such initiatives is found in the results of SERPLAN's 1988 questionnaire survey of its constituent local authorities to assess the quality of the shopping environment in town centres (SERPLAN, 1989a, 1989c). This confirmed that much remains to be done in London. Only 41% of centres had full pedestrianization schemes, with the majority comprising less than 25% of the total shopping frontage. Although it is not possible to make an accurate estimate from the available data (SERPLAN, 1989c), a large part of this modest total must be attributable to new shopping malls. Most centres had only between 0.5 and 3.0 car spaces per 100 square metres of retail space (compared with between 1.7 and 6.0 in the metropolitan ring). Only 17 centres, less than a quarter of the 76 surveyed in London, had benefited from traffic management schemes since 1980, although there were proposals for another 13. New roads had been completed in 13 centres. Fortunately, there is a growing realization of the problems attendant on large-scale new developments in existing centres. Thus, the DOE's 1989 Strategic Planning Advice for London suggests in the preparation of Unitary Development Plans boroughs should give consideration to '. . . the possibility of pedestrianization, to the provision of additional car parking and traffic management measures, and to the importance of public transport' (paragraph 72).

Superstores and retail warehouses

Assisted by a shift in government policy, these new types of shop have had a significant impact on the spatial form of retailing in the London metropolitan region. Superstores[5], mainly developed by the leading supermarket companies, dominated the first phase of decentralization nationally (Schiller, 1986). The second wave involved bulky items – DIY supplies, furniture, carpets and white electrical goods – which was sold from retail warehouses[6]. While the first generation of retail warehouses were accommodated in converted premises, typically on industrial estates, the new generation offer the shopper purpose-built, well-designed stores with good access and car parking provision. Retailers of less bulky comparison goods – such as Marks and Spencer, Toys R Us and Habitat – have in the last few years also been opening out-of-centre stores in a third wave of development, which poses an even greater threat to the well-being of town centres.

London is relatively poorly served by superstores. Davies and Sparks (1989) have recently demonstrated that applications were slow to take-off, and that when they did there was a high rejection rate. Sumner and Davies (1978) attributed the slow growth of superstore applications to the lack of suitable low cost sites and the restrictive attitude of local planning authorities; while LPAC (1987) suggested that the large number of supermarkets may have acted as a deterrent to superstore development. In 1988, London had one superstore for 199 000 people, a lower level of provision than in any other region in the United Kingdom (the rest of the South East had 1:149 000 people). By way of comparison, the best provided region – the East Midlands – had 1:99 000 people (Unit for Retail Planning Information, 1988). However, the number of superstores in London increased dramatically during 1987–88 from 22 to 34, and schemes in the pipeline should enable London to overhaul at least two regions by 1990. In contrast, London is well served by retail warehouses. In 1988, it had 1:60 000 people (only the South West was better served) while the level of provision in the rest of the South East was 1:78 000 (SERPLAN, 1989a).

It is beyond the scope of this paper to investigate in detail the reasons for this relative under- and over-provision. However, to assess the impact of local planning policies on the geography of superstores and retail warehouses, attitudes to superstores and retail warehouses were examined for the 24 boroughs which have Local Plans. The GLDP is silent on out-of-centre developments. However, only two boroughs – Kensington and Westminster – do not have an explicit policy. Elsewhere, three types of approach may be recognized. Three boroughs confine themselves to a general statement of policy on large scale retail developments; nine boroughs include specific policies for retail warehouses in a general policy statement; 10 boroughs have specific policies on both superstores and retail warehouses. All the Local Plans examined were produced in the 1980s, by which time planners were fully aware of these new forms of retailing and had moved away from the negative stances which Gibbs (1987) found were common in earlier years. New developments are universally accepted in or adjacent to town centres; boroughs only differ on whether a set of criteria have to be satisfied for planning permission to be granted. In contrast, the range of attitudes to out-of-centre development varies from outright resistance to acceptance in specific areas or if certain criteria are satisfied. These policies are summarized in 10 categories in Table 7.3; in which category one represents the most restrictive policy, category 10 the least restrictive. Sixteen of the 22 authorities with policies are committed to resisting out-of-centre superstores compared with only two opposing retail warehouses. Almost a quarter are prepared to grant permission to retail warehouses in specific areas or if the development meets explicit criteria.

These findings are consistent with the view that superstore development was initially held back in London by strict planning controls, and that the recent surge in development is a consequence of the central government's desire to make planning less restrictive. Hence, in 1984 the DOE downgraded the importance of Local Plans to just one of the material considerations to be taken into account in determining development applications. Then, in a parliamentary answer by the Secretary of State for the Environment in July 1985, superstores and retail warehouses were effectively accepted as significant features of the shopping scene; and planners were advised that they should not be concerned with competi-

Table 7.3 Local authority policies for new forms of retailing and development outcomes

Policies		Number of local authorities		Development outcomes							
				Superstores 1988		1986			Retail warehouses 1988		
In-centre	Out-of-centre	RW	SS	C	O	C	O	X	C	O	X
1 With criteria	Resist	0	6(5)	5	3	1	1	2.0	1	1	2.0
2 No criteria	Resist	2(1)	10(4)	10	5	0	4	2.0	1	12	6.5
3 With criteria	Resist but with exceptions	4(3)	3(0)	5	2	0	4	1.3	0	13	4.3
4 No criteria	Resist but with exceptions	2(2)	1(0)	0	1	2	13	3.0	3	15	3.6
5 With criteria	With criteria[1]	2(2)	0	0	0	2	22	3.0	6	31	4.6
6 No criteria	With criteria[2]	5(4)	1(0)	2	2						
7 With criteria	Specified areas	0	0								
8 No criteria	Specified areas	2(2)	1(1)								
9 With criteria	Criteria/areas	3(3)	0								
10 No criteria	Criteria/areas	2(2)	0								

Source: *Author's review of Local Plans; Smith (1986) and SERPLAN (1989b)*

Key: RW   Number of authorities with type of policy (specific or implied from general policy) to retail warehouses (the number in parentheses refers to number of authorities with specific policy)
   SS   Number of authorities with type of policy (specific or implied from general policy) to superstores
   C   Number of developments in or adjacent to town centres
   O   Number of developments out-of-centre
   X   Average number of developments per local authority

Notes: 1 Includes outcomes for categories 7 and 9
   2 Includes outcomes for categories 8 and 10

tion. This answer was reproduced and expanded in Planning Policy Guidance 6 (PPG6) on major retail development, which was issued by the DOE in January 1988. Today, the burden of proof to demonstrate that new developments bring harm to interests of material importance now falls clearly on the local planning authority. More specific to London, nine boroughs lost well-publicized appeals on out-of-centre developments in the period 1982–87 (Lee Donaldson Associates, 1986, 1988).

The net result of these policy shifts was that a SERPLAN survey in 1985 found that almost half of the 24 responding London boroughs had experienced problems in operating their retail policies. Two years later virtually all boroughs had encountered difficulties (LPAC, 1987). Local planning authorities have scored one success: they have so far been moderately successful in steering development to town centre or edge-of-centre sites. Thus, 40% of the 53 superstores[7] built or under construction in October 1988 were out-of-centre (Table 7.4), with a third of these won on appeal. However, the balance is shifting rapidly, for 17 months earlier, only one-third were out-of-centre (LPAC, 1987). That being noted, it should also be acknowledged that superstores are not evenly distributed across the London boroughs. Inner London is slightly less well-provided, with one store per 147 000 people compared with one per 117 000 in the suburbs. Two boroughs (Havering and Wandsworth) have four superstores, but eight boroughs await their first. This uneven distribution cannot be attributed to the impact of Local Plan policies. As Table 7.3 illustrates, boroughs have experienced difficulties in operating their retail policies with respect to superstores: eight out of 13 out-of-centre superstores, for example, have been built in boroughs committed to resist such developments.

Table 7.4   The location of superstores and various types of retail warehouse, 1988

	Superstores	DIY	Furniture	Electrical	Others
Town centre	24(51.1)	11(12.0)	2(4.5)	1(12.5)	0(0.0)
Edge-of-centre	4(8.5)	10(10.9)	2(4.5)	0(0.0)	1(12.5)
Out-of-centre	19(40.4)	71(77.2)	40(90.9)	7(87.5)	7(87.5)

Source: Smith (1986) and SERPLAN (1989)

Note: The figures in parentheses are column percentages

Local planning authorities in London have adopted a more relaxed attitude to retail warehouses than superstores so that the new emphases in government policy have been less important. Gibbs (1987) outlines the process by which retail warehouses have gained acceptance nationally. In national terms, at least in the early 1980s, the planning policies of the London boroughs could be considered relatively restrictive (Wilson, 1984). Retail warehouse development in London in 1986 was concentrated in relatively few boroughs (Smith, 1986). Brent (with 21) and Merton (with 10) stood out as important centres; Hillingdon (7), Waltham Forest (7), Greenwich (6), Redbridge (6) and Barking (6) were also prominent. Nine local authorities, seven of these in inner London, did not have any. Developments between 1986 and October 1988[8] by which time at least 172 warehouses were open or being constructed, have made for a more even distribution, although inner London remains poorly served. Six inner London boroughs still await their first, and one warehouse serves an average of 119 000 inner London residents (compared 27 500 people in the suburbs).

The low level of provision in inner London probably reflects lower demand from traders on account of a less affluent population and higher land values more than restrictive planning policies (Smith, 1986). Thus, five inner London boroughs which stand at the liberal end of the policy continuum (categories 5–10 in Table 7.3) had just two retail warehouses between them in 1986, and only five in 1988. As far as can be judged from Local Plans, however, differences in planning policy do appear to be influential in outer London. Of the authorities with specific policies on retail warehouses, the five with the most restrictive (categories 1–4) were served by a total of 10 outlets in 1986, whereas 30 were open in the seven boroughs with more liberal policies. Brent does not have a borough-wide Local Plan, but admits to operating a lax policy in the early years of retail warehouse development (personal communication). With central government pressure for relaxation, the gap had narrowed by 1988, but the more progressive boroughs were still the better provided, with an average of 7.6 retail warehouses per borough compared with 5.0 elsewhere. One other factor which appears to be important in the outer boroughs is the availability of suitable sites. Many traders in the early years, when the attitude of planning authorities to out-of-centre retailing was being tested, sought out warehouse-type buildings on industrial estates with adequate parking and good servicing arrangements (London Borough of Hillingdon, 1982). Four out of five warehouses in London converted or built prior to 1986 were on industrial or mixed industrial/warehouse land. Most probably, therefore, it is not coincidental that boroughs such as Brent, Merton, Waltham Forest and Greenwich, which had large tracts of industrial land ripe for development, are well provided with retail warehouses.

Traders' attitudes to sites have evolved during the 1980s as retail warehouses have become a more accepted feature of the urban scene. Some early warehouses have closed, and other outlets have moved locally to better sites. Many warehouses are still constructed on former industrial land, but sites near major traffic arteries with good access are now sought. Until recently, operators were seeking representation of approximately one store per 50 000 people, but new, higher order operators require a catchment population of at least 100 000 people, and some as many as 500 000 (Drivers Jonas, 1989). Spatial propinquity to other outlets, bringing with it the advantages of cumulative attraction (Nelson, 1958) and spin-off trade from other warehouses, is consequently set to become an ever more important locational factor. The Olympic Estate at Wembley, in the London Borough of Brent, which had nine traders by 1985 (one DIY store, six furniture/carpet stores, one electrical outlet and one unit selling kitchen fittings), is an early example of a cluster which developed through incremental growth. A four mile stretch of the Edgware Road, between Burnt Oak and Cricklewood, and Purley Way in Croydon, are the two main examples of urban arterial growth. Purley Way exemplifies the evolution of this type of strip. The first warehouses (Carpetland and Texas Homecare), constructed in the mid 1970s, were warehouse conversions on the rear of industrial plots with inadequate parking and poor visual prominence. The construction of purpose-built structures with adequate parking commenced in the early 1980s, but at the end of 1985 there were only five units (two DIY, one furniture, one carpet and Halfords). There are now 10 outlets, with planning permission for a further five, along with up to 29 250 square metres of retail floor space in a large scheme which includes housing and leisure elements on the site of the Croydon Power Station. The range of stores has diversified with the opening of Currys and Habitat in 1987, and Childrens World and a Sainsbury's superstore in 1988. Habitat was a direct relocation from the town centre, and the other two durable goods stores are in direct competition with town centre outlets. Outstanding constants include Leatherland, the World of Leather and Toys R Us.

The logical outcome of the need for traders to cluster is the development of planned retail parks. These, as with their stand-alone predecessors, are developed more often than not on converted industrial estates. To date London only has three, at Beckton in Docklands (opened 1986), Edmonton (1988) and Tottenham (1989). Although there are a number of outstanding applications for new developments, London is relatively underprovided with retail parks and also has the lowest floorspace per thousand population in the pipeline (Hillier Parker, 1989). It may be that, as LPAC suggested with superstores, the pressure for development is less because of the amount of floorspace already provided in stand-alone, albeit frequently clustered, outlets.

The number and location of superstores and main types of retail warehouse are listed in Table 7.4. DIY outlets continue to be numerically dominant, followed by furnishing and carpets. Electrical stores are probably under-enumerated on account of their smaller size. To a planning profession largely dedicated to preserving the health of existing centres, Table 7.4 paints a disturbing picture. The majority of superstores, which add substantially to traffic congestion in existing centres on account of the 'weekly shop' and whose decentralization would cause least loss of trade because food shopping is seldom part of a larger shopping expedition, are located in, or on the edge of, town centres. DIY warehouses, a new form of retailing little represented in centres until recently, are the least dispersed of the warehouses. Furniture warehouses, electrical retailers and other traders (including toys, leather goods and motor accessories), which together represent the greatest challenge to the health of existing centres, are more decentralized. And yet, these distributions in part reflect planning policies laid out in Local Plans, while LPAC (1988, paragraph 5.19) still places particular emphasis on large convenience foodstores being in or adjacent to larger centres. It is as though planners failed to realize the full implications of retail warehouses until the momentum for their development was unstoppable. Clusters of retail warehouses, whether planned or unplanned, comprise a totally new element in the retail scene, with a new function which does not fit neatly into the concept of a retail hierarchy. Their impact on traditional centres is not yet clear.

A report commissioned by the London Borough of Croydon, has recently been completed on the impact of the complex of retail warehouses on Purley Way on the centre of Croydon, little more than a mile away (Drivers Jonas, 1989). The work was bedevilled by lack of turnover data, but the consultants estimated that spending in retail warehouses in the borough in 1986 was approximately £20 million, about two-thirds of which was generated within the borough. This local expenditure of £13 million almost exactly matches the increase in local expenditure on furniture, floorcoverings, household textiles, hardware and DIY supplies between 1976 and 1986. Nonetheless, DIY, furniture and carpet outlets in the town centre had decreased. The town centre, with an estimated turnover of £331 million, had adjusted with a new mix of traders and continued to trade well. However, there was some evidence of decline in the secondary shopping zone towards the edge of the centre. Other centres have been less fortunate. LPAC (1987) reported that individual out-of-centre superstores have led to modest town-centre schemes in Woolwich and Harlesden falling through. Insofar as it is possible to draw conclusions from scant evidence, it appears that superstores and retail warehouses, even when clustered, have been absorbed into the retail system with a minimum of dislocation in areas where town centres are thriving, but that they threaten to add a further twist to the downward spiral of declining centres.

Regional Shopping Centres

No discussion of retailing in London at the turn of the decade would be complete without a discussion of regional shopping centres, although the first will only open in 1990. There have been 16 applications since 1986 for regional shopping centres within the area defined by the 30 minute isochrone from the M25. Eleven of these are located in its immediate corridor. The reasons for this intense development pressure are well documented. Nationally, a period of rapid growth in expenditure has coincided with a perception among developers that regional shopping centres were at last fair game in the national policy vacuum that existed with respect to major retail developments prior to the publication of PPG6. Regionally, the completion of the M25, with 4.5 million people living within a 10 minutes drive, has made for a number of highly accessible sites at interchanges. It has been estimated, for example, that 3 million people live within 30 minutes drive time of the proposed regional shopping centre at Waterdale Park, near St Albans. These nodal sites are situated in a very affluent region, with one in four households outside London having two or more cars.

It is still too early to predict with any degree of certainty the distribution of regional shopping centre's in the London region at the turn of the century. The government has still to respond to requests to issue strategic guidance for their development. Each case is taken on its merits in the light of PPG6 and, although they do not specifically address retail development, in the context of letters of guidance to SERPLAN, which the DOE brought together in 1988 in its Regional Guidance for the South East (PPG9). In the absence of stronger policy direction, our interpretation of current official thinking must be based on examining those applications which went to appeal. Although each development has its unique aspects, there appear to have been four determining factors: quantitative and qualitative need; impact on other shopping centres; impact on the M25; green belt policy.

Quantitative and qualitative need

In constructing their regional guidance to member authorities, SERPLAN examined the quantitative and qualitative needs for regional shopping centres. They concluded that over three-quarters of the estimated increase in floorspace needed between 1984 and 1996 is already in the 'pipeline', with applications already lodged for half as much again. It concluded, therefore, that there was no quantitative need for regional shopping centres, although there was likely to be a shortfall of floorspace in the corridor between the M3 and M40. A qualitative need is considered to exist if shoppers are not within reasonable reach of a full range of shopping in a high quality environment. SERPLAN considered there to be qualitative gaps in shopping provision in three corridors: between the A10 and M11, between the A12 and A13, and the A2 corridor in North Kent. While these analyses seem redundant in the light of DOE guidance that it is not the role of planning to prescribe floorspace limits, developers have been quick to use them when it is in their interest. However, no Inspector has viewed qualitative or quantitative needs sufficiently important to override other factors: they have been variously viewed as 'arguable', 'not established', 'not pressing' and 'not providing decisive support'.

Impact on other shopping centres

Much of SERPLAN and LPAC's concern about regional shopping centres flows from their potential impact on existing shopping centres. While these fears may be well grounded if a large number of schemes were to be completed, impact on existing facilities has not been a determining factor in any of the appeal decisions, and not even a major point of contention between the parties in two public inquiries.[9] Although one or two individual centres in close proximity may suffer, in general the impact of an *individual* regional shopping centre has been viewed as diffuse.

The M25

Traffic congestion on sections of the M25 is clearly going to be a determining factor in decisions in parts of the region. The M25 is already carrying up to 50% more cars than what is considered a reasonable daily maxima in certain sections, notably between junctions 10 and 21. The local authorities have estimated that Wraysbury regional shopping centre would generate 23 000 vehicles per day; and with general traffic growth to 1992, the total flow in that year between junctions 13 and 14 could reach 196 000 per day on a dual four-lane road designed for 105 000. The Department of Transport directed the Royal Borough of Windsor and Maidenhead to refuse Wraysbury regional shopping centre on the grounds, *inter alia,* of the additional traffic being detrimental to the free flow of traffic. The developer's appeal was dismissed, in part, on the grounds of motorway congestion.

Green belt policy

Of all determining factors, green belt policy appears to have been easily the most influential. PPG6 states that major out-of-town developments '. . . have no place in the Green Belt' (paragraph 15). This view is reaffirmed in the DOE's Strategic Planning Advice for London. The four regional shopping centres in the green belt for which decisions have been handed down have all been rejected, although one site comprised disused mineral workings and another was surrounded by two motorways and a trunk road. Their undermining of green belt objectives was an important determining factor in each case.

The two large centres which have received permission do not impinge on green belt land. Permission has been granted on appeal to a development of 107 000 square metres of gross retail floorspace along with a multiscreen cinema at Lakeside, Thurrock, near the entrance to the Dartford tunnel. Retail parks comprising 72 500 square metres of retail floorspace are already trading nearby. Several special factors influenced this decision: the area is economically depressed, and the development was supported by the Thurrock District and by Essex County Council. The second approval is for an in-town development of 78 500 square metres at the Royal Albert Dock in Docklands on Development Corporation land.

The picture of likely provision in the M25 corridor in the mid-1990s is becoming clear. In February 1990, of the 10 applications in addition to Thurrock, five have been withdrawn, four have been refused on appeal (Runnymede Centre, Wraysbury; Hewitts Farm, Orpington; Waterdale Park, St Albans; Aldenham

Retail Park, Harrow). The appeal decision is awaited on Blue Water Park, Dartford. Although this is zoned as green belt, the site is an abandoned chalk quarry and Kent County Council has been seeking to change its designation in its new structure plan. This has delayed the appeal decision, but the Secretary of State for the Environment has announced that he is minded to grant planning permission. If this proposal does finally get the go-ahead then there will be three regional shopping centres in a narrow arc to the east of the centre of London, and none elsewhere in the M25 corridor; which is a sad commentary on the state of regional planning in the metropolitan region. It is difficult to predict decisions on the two in-town developments at White City and Hackney Wick, but they may fall on impact grounds.

The Future in Greater London

Planning in Greater London entered a new era in July 1989 with the publication by the DOE of the Strategic Planning Advice for London. Boroughs are now beginning to prepare Unitary Development Plans within the framework of this guidance. LPAC (1988) prepared a detailed submission to the DOE on the retailing issues it believed this advice should address. They requested that a broad framework be developed to cover four topics: the number of, and broad locations for, regional shopping centres in the M25 corridor; proposals and criteria for the maintenance and forward planning of the existing hierarchy of strategic centres; proposals and criteria for out-of-centre shopping; proposals to enhance the role and function of the West End and Knightsbridge. The ensuing DOE guidance devotes 277 words to retailing. Fifty-three of these are reproduced from PPG6; the only new material is advice on measures to improve town centres. It makes no attempt to address London as a unique place.

This observer finds both the LPAC submission and the DOE guidance profoundly depressing. LPAC is still looking back to the 1960s rather than forward to the 21st century. It appears to have learned nothing from the failures of the GLDP: it proposes 12 strategic centres in inner London, including Brixton and Peckham, without any positive proposals about how to revitalize those that are in steep decline. LPAC grudgingly accepts there may be circumstances in which out-of-centre developments may be acceptable (if they meet eight criteria) but new shopping developments should wherever feasible take place in existing centres. It is as though local planners are trying to 'fossilize' the retail system in a form that evolved when public transport reigned supreme. On the other hand, the DOE strategic advice is 'about as useful as treating a broken limb with cotton wool' (Royal Town Planning Institute President Chris Sheppey, quoted in *The Planner*, August 1989). It is difficult to see how 33 London authorities are going to coordinate their retail planning in the absence of proper strategic advice. There surely must be a middle way by which a modern retail system with a new form suitable to the 1990s, sensitive to the needs of both car-owners and those who rely on public transport, can be developed in the metropolitan region by developers working with forward-looking local planning authorities in the context of a detailed strategic plan. At the moment this looks to be a pipedream.

Notes

1   SERPLAN do not divide inner London into sub-areas. Their analysis is consequently based on 11 areas.

2   Centres with 10 or more multiples are included, following Schiller and Jarrett (1985).

3   The author was also concerned that, given the propensity of chain stores to cluster in new malls, the Schiller-Jarrett index would favour centres with newly constructed floorspace. However, there was no evidence of this from a comparison of this index with a more composite index (SERPLAN, 1987).

4   Threshold is the minimum amount of purchasing power necessary for a retailer to make a profit by selling a good.

5   Superstores (the larger of which are termed hypermarkets) are self-service establishments with a sales area of at least 2500 square metres selling a wide range of merchandise but in which foodstuffs represent more than one half of sales.

6   Retail warehouses, which represent an extension of the wholesale cash and carry trade, sell non-food goods directly to the public. They are usually single storey structures with adjacent surface parking. They tend to range in size from 1000 square metres to 8000 square metres.

7   This data base updates the Unit for Retail Planning Information (1988) list of UK hypermarkets and superstores with information from the SERPLAN (1989b) survey of retail developments of October 1988.

8   This data base updates the survey by Smith (1986) with information from SERPLAN (1989b). Both these surveys were directed towards developments in excess of 2300 square metres, although they do include smaller developments. However, it is not clear how many developments of less than 2300 square metres, which are more likely to be selling electrical items and goods for motorists, are excluded.

9   Adanac Park regional shopping centre, Southampton, which is outside the study area, was refused on appeal partly because of its expected impact on Southampton and Eastleigh.

8

Transport: How Much Can London Take?

Tim Pharoah

London reached its maximum size before the car appeared on its streets in great numbers. The growth of London to more than seven million people by 1920 (at which time it was the largest city in the world) was made possible by the building of a dense and comprehensive railway network (again unrivalled in the world). Since then, London has absorbed two and a half million motor vehicles, and more are arriving at the rate of one hundred a day. In this chapter, by examining transport trends within the London region, we shall attempt to assess this transition from the railway age to the motor age.

London: A Tale of Two Cities

The functional and transport structure of London may be presented as two cities in one. The first consists of its large and functionally highly specialized core, accessible via a network of railways radiating from the centre. The second is the vast collection of suburbs which to an important degree have a life of their own, based not on rail but on road transport.

The 'first' London

The *raison d'être* of our first London is the commercial and cultural core area of some sixteen square kilometres where about 1.2 million people work, roughly one third of all Greater London employment. Less than 15% of these workers actually live there, so each weekday over one million people commute into central London from outside. Most of these journeys are made within two two-hour periods each day. This spatial and temporal concentration of movement is possible only because three quarters of the demand is met by rail transport (Table 8.1). No other mode of transport can perform such a task on such a scale, and consequently the central area is dependent on rail for its very existence. This dependence is reinforced by the fact that central London workers live further from their work than do workers in inner and outer London, as Table 8.2 shows.

If London is to maintain this high concentration of activity at the centre, then there is no alternnative to maintaining the rail system. Although this may require heavy long-term public expenditure on fare subsidies and capital investment, it may still be regarded as an efficient and popular way of organizing transport in the

Table 8.1   Commuting to central London, morning peak 0700–1000 hours

Mode	Passengers in thousands	% passengers
British Rail	468	40
London Underground	411	36
(Total Rail)	(879)	(76)
London Bus	80	7
Commuter Coach	21	2
(Total Road Public Transport)	(101)	(9)
Private Car	160	14
Motor/Pedal Cycle	17	2
(Total Private Transport)	(177)	(15)
Total all Modes	1157	100

Source: London Regional Transport, 1989

capital. In any case, as Buchanan demonstrated, restructuring the city centre to accommodate a high proportion of journeys by car is neither physically nor financially feasible (Buchanan, 1963, 141).

The growth of central London has arisen from its position as the most accessible place in the country. This assertion may at first seem to be at odds with the acknowledged costs and problems of commuting to central London and congestion within it. But it is important to recognize that although the average cost of reaching the centre is higher for a Londoner than for a resident of a smaller city (in terms of time, money and other costs), the generalized cost of bringing together well in excess of one million people each day is lower for central London than for any other location in the United Kingdom. Thus, in *relative* terms, central London is the most accessible location by virtue of the fact that the main railway routes focus upon it, and also that 90% of the Greater London population lives within one mile of a station providing direct services to it (Greater London Council/Department of the Environment, 1974, 22).

For certain activities, the advantages of a central London location clearly outweigh the disadvantages (of high travel costs and high land and rent costs); otherwise the level of central London employment would not have grown or remained so high. But the advantages are not valuable to *all* employment, and

Table 8.2   Daily trips by Greater London residents, 1981

	No. of trips (thousands)	% trips	Distance travelled (thousands kilometres)	% distance
Walk	7149	36	7792	7.3
Pedal Cycle	458	2.5	1265	1.2
Car/Van	7866	40	52516	49.5
Motor Cycle	224	1	1496	1.4
Bus	2252	11.5	13262	12.5
Underground	941	5	10662	10.0
British Rail	634	3	13962	13.2
Other	172	1	5199	4.9
Total	19696	100.0	106154	100.0

Source: Greater London Council (1985a)

the balance of advantage and disadvantage varies for different kinds of business, and may change over time. Thus, whilst some firms or parts of firms decide to move away from central London, others move to the centre or start businesses there. This process of outward movement and replenishment is probably as old as London itself, but over the centuries the activities at the centre have become more and more specialized. As a residential area it has declined steadily, manufacturing industry has all but disappeared, and general office functions have been dispersing to the suburban areas and beyond. The need to pay higher salaries in London is a further incentive to decentralize when a central location has no particular advantage.

The key feature of central area activities is the extensive geographical coverage of their linkage patterns, usually because of their need either for specialized labour or for specialized customers. This applies, for example, to international banks, fashion houses, and opera houses, as well as to offices, which are the main source of central area employment. Major office growth in the 1960s was followed by decline in the 1970s. During the 1980s, growth in the finance, insurance and banking sector appears to have succeeded in reversing the decline in central London employment. For example, employment in this sector increased by 8% between 1981 and 1984, compared with losses of 10–15% in most other sectors including manufacturing, transport and construction (see Chapter 3). Jobs in the declining sectors have tended to be located in the suburbs, whereas those in finance are concentrated at the centre (Greater London Council, 1985b).

Thus the daily surge of 800 000 people commuting by rail is a dramatic reminder of the scale and power of London's traditional core. The strength of the centre depends upon the railways that link it to the rest of the city, and the centre is also the *raison d'être* of the railway system. The railways perform three principal roles in our 'first' London. First and foremost is its dominance of peak hour commuting to the centre (Table 8.1) and distributing passengers within the centre (mainly Underground). Secondly, the railways link the centre of the capital with other cities and towns, and with other countries via the major London airports and the ferry ports on the South and East coasts. This function will be greatly heightened by planned new rail links to Heathrow and Stanstead airports, and by new rail services to continental Europe through the Channel Tunnel. The third main role is catering for non-work and off-peak journeys throughout the South East region; although despite increases during the 1980s, these form only a quarter of British Rail's Network South East and one-third of Underground journeys. Taken together, these passenger and freight services comprise the world's largest concentration of railway activity.

We have thus outlined our 'first' London, one which is structured around a high capacity radial railway system converging on a dominant and specialized core of activity, producing a so-called 'hub and spoke' pattern.

The 'second' London

London's enormous railway enterprise nevertheless accounts for a fairly small part of the capital's total travel picture. There is a 'second' London which is not so closely bound up with the railway network. Taking journeys by London residents as a whole, in 1981 rail accounted for 16% of journeys by motorized means, and only 8% of total trips when walk and cycle trips are included (Table 8.2). In terms of distance travelled, about 20 billion passenger kilometres are carried annually on British Rail's Network South East plus the Underground, but the roads of the much smaller Greater London area carry about 43 billion passenger kilometres. Even in terms of travel to central London, it is road and not rail that is the dominant mode over the 24 hour day, with rail carrying roughly one quarter of total trips (Department of Transport, 1988). Rail accounts for a small proportion of total freight movement in London, probably less than 5% of tonne-kilometres. Looking at the different journeys which Londoners undertake, rail is used as the main mode for 20% of work trips (mostly to the centre), but for only 6% of education trips and 2% of shopping trips.

Apart from its dependence on road travel, the essential feature of our 'second' London is that the suburban (non-central) parts of the city are less specialized in their function than the centre itself. The shops, schools, hospitals, clinics and workplaces are not noticeably different, in terms of the role that they perform, from one London suburb to another. Many of these facilities are grouped into identifiable sub-centres which serve their particular catchments of local population rather than performing a wider metropolitan function. This is not to say that London's suburbs are a uniform, undifferentiated sprawl. Indeed, the first Draft Strategic Guidance for London states that 'One of London's particular strengths is the distinctive identity and character of the many localities and communities' (Department of the Environment, 1989, 2). While the central area depends on rail commuting, the suburban centres with their relatively small catchment areas, rely more on car, bus and other modes better suited to local journeys.

The presence of retail, commercial, employment and other activities in suburban centres is an important feature because it enables people to gain access to most facilities (especially the non-specialized ones) without travelling long distances. To the extent that such facilities are grouped in town centres, it also enables people to reach them without being dependent on the car. A single bus journey will bring most suburban Londoners to a centre which will cater for their daily requirements. All the suburban centres have bus routes which focus on them and which serve their catchment populations (on average about 200 000 people). It is therefore still true to say that if, for some reason, all private cars disappeared, London could continue to function without major difficulty. All that would be required would be for people to walk a little more, for extra buses to be provided, and for people to dust off the bicycles that have laid unused for so many years. This may be a little over-stated, but it serves to illustrate an important feature of London which distinguishes it from motor-age cities such as Los Angeles or Houston.

The localized nature of much of London life is recognized by Londoners themselves. It is common for residents of Harrow or Bromley to talk of going 'up to town', or even 'up to London' for the theatre or for Christmas shopping. Most people in London shop locally, go to school locally, socialize with others who live locally, and make use of a whole range of medical, professional, financial, building and other services provided in their locality.

This, then, is our 'second' London. It is a realm of substantially self-contained suburban communities. Taken together, these two different but interdependent aspects of London comprise the capital's traditional land use and transport structure which was established before the motor age.

The Impact of Motorization

We now turn to the question of what has been happening to London's land use and transport structure in the years since the Second World War. From a transport point of view, the most precipitous change in London as elsewhere has been the increasing dominance of the car and the lorry. Two and a half million vehicles have now invaded London, transforming the lives of its inhabitants and setting in train fundamental changes to the fabric and appearance of the city. Since the early 1950s, when their ownership came within the financial reach of middle-income households, motor cars have come to dominate the entire metropolitan environment and, as shown in Table 8.3, to become the major form of personal travel. This increase continues today. In the 10 years from 1978 to 1988 an average of 100 additional cars rolled onto London's streets every day. A large proportion of these went to households already in possession of at least one car.

Table 8.3   Daily increases in radial road traffic 1983–87

	Greater London cordon %	Inner London cordon %	Central London cordon %
Private cars	+ 9	+ 5	+ 1
All vehicles	+ 8	+ 3	+ 0.4

Source: Department of Transport (1988, 8)

Car ownership is not evenly spread across the London region, however, and may be broadly described as increasing with distance from the centre. This reflects not only variations in prosperity (of residents) but also the availability of parking and road space, both of which become more plentiful with distance from the centre. Also the quality of alternative means of transport declines with distance from the centre thus providing a stronger incentive for car ownership in outer London, especially multiple car ownership.

This increasing motorization has destabilized the traditional 'dual structure' of London we have described, and created a range of difficult policy issues concerning the provision and maintenance of transport capacity, the management of transport demand, and changes to the city's land use structure. To judge the impact of the car on London's structure, let us consider two extreme models of land use/transport structure. The first is a high density structure with diverse and numerous activities, served predominantly by walking and public transport. This may be regarded as the model which best describes the central parts of many if not most European cities. The second is of a loose, low density structure without large concentrations of activity or recognizable 'town centres'; where most travel is undertaken by car, where the diversity of trip patterns precludes high capacity public transport services, and where high average trip lengths preclude walking as a major mode of travel. The interesting feature of these two extremes is that while each can work in its own terms, it is in almost every way incompatible with the other (for further discussion of such urban 'archetypes' see Thomson, 1977). The innermost parts of London can still lay claim to being an example of the first of these extremes. The second 'car-based' extreme can be found in parts of outer London, and settlements in the outer metropolitan area, although the more profound examples occur in North America.

The question arises as to what happens where these two incompatible urban forms meet, as they inevitably must? The answer, broadly speaking, is a more or less unsatisfactory compromise. The conflict between London's pre-car physical structure and the growing forces of motorization is expressed in deteriorating environmental conditions, chronic road congestion and a declining public transport system. Dependence on the car has a firm grip in the outer parts of the London region, but not at the centre. The crucial question is how far in towards the centre could or should motorization spread, and what are the implications for developing the suburbs, and maintaining the rail-dependent city centre? The following paragraphs examine this fundamental issue, and the trends that are relevant in resolving it.

Mobility versus access?

The value of increased personal mobility brought about by the motor car has been challenged by Hillman, Plowden and others who argue that what is important is access of facilities, not movement *per se,* and that increased opportunities for car owners have been provided at the expense of declining opportunities for others (see Hillman *et al.,* 1976; Plowden, 1980).

It is often assumed that 'greater mobility' brought by the car involves people making more journeys. The argument is that as higher incomes and shorter working hours provide more 'disposable time' so people undertake more journeys in order to undertake more activities. In London, however, there is no firm evidence that the number of journeys undertaken by residents has increased since the Second World War. The number of *motorized* journeys has increased (by 7% between 1962 and 1981) but much of this has simply offset a decline in bicycle use and perhaps walking.

What has certainly occurred is a dramatic change in the mode of travel, namely massive growth of car trips at the expense of bus, bicycle and possibly walking trips. Also, there has been a large increase in total travel (as measured by passenger and vehicle kilometres) associated not only with the shift to private motor transport, but also with the outward movement of population. Journeys increase in length with the distance of journey origin from the centre. For example in 1981 journeys within the central area were on average 1.25 kilometres in length, in inner London 2.34 kilometres and in outer London 3.33 kilometres (Greater London Council, 1985a).

The most visible impact of motorization, of course, has been the increased traffic on London's roads. Estimates based on the major transport surveys show vehicle kilometres doubling between 1962 and 1981, from 39 to 80 million kilometres per day (Thomson, 1969, 28; Greater London Council 1985a, foreword). Between 1983 and 1987 traffic is estimated to have increased by 7% (Department of Transport, 1989a). However, traffic growth, like that of car ownership, has been higher in outer London than inner London. In central London there have been only small increases since the mid-1960s (e.g. Table 8.3).

The growth of road traffic, then, has come about as people switch to the car for an increasing proportion of their journeys, as they travel further to carry out their business and, possibly, as they make more trips. Mobility for car users has thus increased, but whether accessibility has improved is debatable. For non-car users accessibility has almost certainly declined because of the associated overall decline of public transport and other modes. The benefits of motorization are therefore not at all clear, unlike its disbenefits which are daily becoming more apparent.

The 'livability' of London

Transport, probably more than any other industry, imposes external costs in the form of environmental damage. The trends towards motorization already described have inevitably and inexorably degraded the quality of London's environment and 'livability'. The noise and fumes produced by 80 million vehicle kilometres every day in London are a major irritant. These traffic levels have also made walking and cycling hazardous and unpleasant. Hillman and associates argued that '. . . the growth of traffic can create conditions in which children in particular sustain a loss of freedom or independence, as parents increasingly feel obliged to impose restrictions on them' (Hillman *et al.*, 1976, 164). The loss of the bicycle and the bus as major modes of travel has in particular restricted the freedom of children.

Heavy traffic flows have split communities throughout London, and are visually intrusive. Many streets have suffered from traffic management schemes (one-way systems and the like) designed to squeeze more capacity from the road system, but which make local movement and access more difficult. Parked vehicles are in evidence everywhere, with many residential roads lined either side with parked cars most of the time. Typically, at least 40% of cars are parked at home even at the busiest times of the week (Pharoah, 1986). The toll of human misery brought about by road traffic finds dramatic expression in road accidents, which in 1987 claimed 456 lives and 48 998 injuries in London (Department of Transport, 1988).

Many other European cities have attempted to change transport priorities to encourage the 'soft modes' of travel (walking and cycling), as well as public transport, and to discourage and mitigate the effects of car traffic. London was an early convert to the concept of traffic restraint, with the introduction of controls on parking provision and use in the 1960s, and strong resistance to demands for major provision of roads and car parking spaces. But measures to protect the environment and vulnerable road users have been introduced only slowly and in a very patchy way, though the exclusion of cars from Oxford Street, and the pedestrianization of some suburban centres has been successful in commercial as well as environmental terms. Indeed, it could be argued that better shopping conditions are essential if the established centres are to compete with the newer out-of-centre pedestrian shopping malls. Traffic management measures have been introduced in some residential areas to reduce 'rat-run' traffic, but London is well behind other European cities in curbing excessive speeds, or introducing what are now known as 'traffic calming' measures (see Pharoah and Russell, 1989). There is considerable scope for making the majority of London's streets more 'livable', but this will involve a change of priorities, and an associated diversion of investment.

Another trend that is especially worrying in London is the distortion of the transport picture created by crime and fear of crime in the streets and on public transport. A survey of women by the Greater London Council in 1985 found that threats to personal safety are a major concern when travelling around London. 'Fears for their safety are so great that 63% of women say that they avoid going out on their own at night' (Greater London Council, 1985c, volume 2, 11). A Department of Transport report also acknowledged that '. . . fear of crime is a significant inhibiting factor to travel on the Underground at certain times' (Department of Transport, 1986, 19). As car ownership increases, public transport will in any case find it harder to compete for off-peak social and leisure journeys. However, fear of crime is likely to cause further shrinkage of this already fragile sector of the public transport market.

Central area traffic restraint

Rail commuting to the centre, and thus the 'first' London described earlier, has been largely protected from direct competition from the car by the limited provision of road and parking space, and by what may be termed 'demand management'. On-street parking has been rationed by parking meters and 'yellow line' waiting restrictions since 1959 (and for new buildings maximum standards of off-street parking apply, unlike the more usual minimum standards elsewhere). As far as road building is concerned, even the most virulent proposals have avoided central London. In 1969, for example, the draft Greater London Development Plan proposed three orbital motorways plus connecting radial motorways in London, but none of these came nearer than 5 kilometres from Charing Cross or St Paul's. In 1973 the two inner orbital motorway proposals were abandoned as being too expensive, too destructive, and counterproductive in terms of traffic congestion. Only the outer ring motorway went ahead (now the M25) though the Department of Transport has continued to expand its trunk road network in the outer suburbs. The difficulty of building roads increases exponentially as one approaches the centre, largely as a result of higher land and construction costs. It has never been possible to demonstrate that the benefits of new high-capacity roads in inner London would outweigh these costs, let alone the environmental upheaval. Central London has thus kept the car at bay by a mixture of deliberate and *de facto* traffic restraint.

The centre has nevertheless continued to gain strength without making concessions to private transport, and further large increases in office employment have been predicted (Department of Transport, 1989b, 5), at the same time as tougher traffic restraint measures such as road pricing have officially been considered.

The Captive Rail Commuter

Commuter rail services will therefore continue to play their highly specialized role in maintaining the vitality of London's economy. The provision of rail services of adequate capacity and quality has, however, become increasingly difficult. Overcrowding has become acute on many lines, while there is a widespread dissatisfaction with standards of reliability, cleanliness and safety. At the same time, fare levels have increased faster than the rate of inflation. The poor standard of service is the result of several factors. Firstly, the rise in commuting (25% from 1983 to 1988) was largely unforeseen and therefore not planned for (Department of Transport, 1989b, 3). Secondly, congested roads have meant that rail has had to shoulder almost the entire burden of 100 000 extra daily commuters. Thirdly, revenue subsidies which would have enabled more frequent services to be run have been reduced to a fraction of their pre-1985 value (Department of Transport, 1988, 26). Fourthly, the revenue base has shrunk as off-peak demand for suburban rail travel has fallen. These related factors deserve closer attention.

Overcrowding cannot generally be attributed to lack of system capacity. By 1987, rail commuting, after a decline in the 1970s, had climbed back to 95% of its 1962 level. In theory, rail capacity should have increased with more efficient and spacious trains on some lines, better signalling, and the addition of some new lines, such as the Victoria and Jubilee lines in 1969 and 1977, and British Rail's Thameslink in 1987. Overcrowding must, therefore, be due to some combination of more peaked journey patterns or less trains operated at peak times. The latter is certainly true of Network South East lines, on which less trains are operated than 30 years ago. Two examples will illustrate the point. Peak hour trains from Streatham Hill (in inner-south London) to Victoria reduced from 22 in 1959 to 10 in 1989 (a drop of 55%). During peak hours on this service the average gap between one train and the next (i.e. headway) increased by 100% (from 8 to 16 minutes). Of course, some decline in service might be expected in inner London where population has also declined. Our second example is Surbiton in outer London where population has been more stable. Taking again the 7–10am weekday period, the number of trains to Waterloo dropped from 40 in 1959 to 16 in 1989 (a decline of 60%).

Reduced services, then, are a major factor in commuter overcrowding, but they are themselves a response to a deeper problem of railway costs. Like many specialized goods and services, the supply costs of commuter railways are high, mainly because of the excessive peaks of demand. Rolling stock, track, stations and staff have to provide huge capacity for two short periods of each weekday, but these facilities are greatly under-used for 80% of the time. It is here that the car has made its presence felt, for it has eroded much of the off-peak demand for rail travel, especially in outer London. For example, Tattenham Corner on the southern fringes of the built-up area had 31 trains each way on Sundays in 1959, but only 15 each way in 1989. Nearby Epsom Downs enjoyed 63 trains each way in 1959, but had no Sunday service at all in 1989.

Such decimation of off-peak rail travel has narrowed the role played by the railways in London, and thus their revenue base. This could, of course, be counteracted by providing revenue subsidies, and this was the policy for many years up to 1985. But subsidies to both London Underground and Network South East have been reduced over precisely the same period that commuting has increased, hence the need for cost cutting, reduced services and higher fares. Here London stands in stark contrast to other European cities such as Paris and Rome where the balance has been struck in favour of high subsidy and service improvements.

Capital investment has tended to attract more political support than revenue subsidies, and London's railways may be on the brink of a new wave of expansion. The Docklands Light Railway, which opened in 1987, was one of the first to be built in Britain, and provided a considerable boost to the regeneration of Docklands. Further large scale growth of employment and population in Docklands will be served by extensions of the Light Railway, and by the Jubilee line extension through Canary Wharf to Stratford. Both these projects have been approved only on account of substantial private sector funding, and are designed to cater primarily for new traffic. Other rail schemes, including new 'crossrail' links between British Rail termini, which are designed to relieve overcrowding, have little potential for attracting private capital, and by the end of the 1980s no commitment to building them has been given[1]. Even though major upgrading of some Underground and British Rail lines did receive government approval in 1988 and 1989, a substantial proportion of costs is to be met from fare revenues.

Particular difficulties remain for Network South East services within Greater London. For many years British Rail has favoured the more profitable long distance commuter services for improvement and investment. Services within the Greater London area have taken the brunt of service cuts, cancellations and limited investment. Of the Network South East investment listed in a Department of Transport statement of 1989 only 3% was specifically for inner London services, 40% would benefit both inner and outer London passengers, and 57% would benefit passengers primarily from beyond the Greater London boundary (Department of Transport, 1989a).

The Death of the Bus?

Rising incomes lead not only to increased mobility, but to people opting for the car in preference to other modes. Thus the bus and the bicycle (and probably travel on foot) have been major casualties of London's growing prosperity. In the early 1950s people used buses for 40% of their trips, but by 1981 buses served only 18% of trips. In the half century from the immediate pre-war period, public road transport services and passenger journeys by road declined in London by 60%.

We have already referred to the strain placed on commuter railways by the switch of passengers from buses. Between 1960 and 1988 both the number of bus passengers and the number of buses entering central London in the morning peak period declined by more than half (London Transport, 1970; Department of Transport, 1988). This downward trend in bus use has meant a substantial decline in the efficiency of the roads as passenger carrying arteries. For example, between 1971 and 1988, despite an increase of 10% in the number of passenger vehicles entering Central London in the morning peak, the number of persons carried by these passenger vehicles declined by 16%. As a result, road use efficiency, measured as the average number of persons per vehicle, declined by 24%, from 2.59 in 1971 to 1.97 in 1988 (Table 8.4). Moreover, bus journeys are slow and unreliable with door-to-door journey times increasing by 21% between 1971 and 1981 (Greater London Council, 1985a). People choose the bus less because it offers a poorer quality service as conditions on the roads have worsened, but also because journey lengths have increased, thus making the bus a less suitable transport mode. As on the railways, the loss of bus passengers has led to a contraction of services. The London bus system has been in retreat for almost the entire post-war period, though it has held up better in central and inner areas and at peak hours than in outer London and at off-peak times.

Table 8.4   Commuting to central London by road, 1960–88 (0700–1000 Hours)

Passengers	1960	1971	1981	1988	% change 1971–88
Bus	215600	156000	121000	101000	
Car	–	163000	173000	160000	
M/P Cycle	–	12000	26000	17000	
Total	–	331000*	320000*	278000	– 16
Vehicles					
Bus	5200	3800	2600	2500	
Car	–	112600	131000	121800	
M/P Cycle	–	11300	26000	17000	
Total	–	127700*	159600*	141300	+ 10.6
Road efficiency as a passenger carrier. i.e. average passengers per vehicle		2.59	2.00	1.97	– 23.9

Source: Letter from London Regional Transport, August, 1989 and LRT (1989)

* Excluding Coach/Minibus

One of the major problems for London's buses is that they have never been fully integrated with the rail system, neither have their operations been restructured to minimize the effects of increasingly congested roads. Some improvements have occurred in the provision of bus and rail interchange (for example at Harrow), and there are about 250 bus priority measures and some bus-only roads (for example at Clapham Junction). Travelcards introduced in 1985 also have enabled passengers to choose between bus and Underground without financial penalty, though in 1989 35% of passengers still bought tickets for individual journeys (London Regional Transport, 1989). Planning and investment have been patchy, and more often than not designed to cope with falling demand rather than to exploit the potential role of the bus. For example, larger buses were introduced so that reduced service frequencies could save on operating costs while still being able to handle peak hour demand. The abolition of conductors on many routes has caused delays at bus stops and is widely disliked.

Rapid changes towards out-of-centre facilities in the suburbs will reduce even further the relevance of the bus network, which is unsuited to serving dispersed patterns of activity, especially where its share of the generated travel is small. The prospects for the suburban bus are therefore bleak in the extreme.

Motorization of the Suburbs

With some exceptions (such as Thamesmead and New Addington) London's suburbs were not specifically laid out to accommodate mass car ownership and use. In the inner suburbs, the car has been accommodated only with great difficulty, and congestion in the ring around central London, sometimes called the 'gluepot ring', can be more severe than in the centre itself. The parking problem, especially in the streets of Victorian and Edwardian terraced housing has become severe. In many areas the conversion of larger older houses into flats, while helping to meet the housing demand from a growing number of smaller households, has generated excessive demand for kerbside parking. Residents complain that they cannot park near their home, and businesses complain of lack of space for deliveries.

Similar problems can also be found in outer London, though they tend to be not so widespread, and more space has usually been made available for off-street parking in what were once front gardens, vacant plots, grass verges or cycle tracks. Particular parking problems occur around railway stations where commuters leave their cars and travel on to central London by train. Parking pressures have also developed around facilities such as hospitals and sports centres. The tendency for such facilities to become less numerous and larger has generated the need for extra travel, a high proportion of which is by car. Overall, the supremacy of the car in outer London is clear from Table 8.5 which compares the mode of travel for work and other journeys in inner and outer London.

The high proportion of journeys wholly on foot is vulnerable if journey lengths continue to increase (for example due to further contraction of local shopping and other facilities). London's established centres have managed so far to retain a large share of retail activity, and to cater for people without access to a car. In 1981 walking was still the main means of reaching suburban centres for shopping purposes, even in outer London, and the car had barely overtaken public transport.

Nevertheless, the retail revolution has not completely passed London by and an increasing proportion of shopping is now located in developments which

Table 8.5   Travel by car and other modes in London suburbs, 1981

Mode of travel	Inner London To Work 1981 %	(1954) %	Other Trips 1981 %	Outer London To Work 1981 %	(1954) %	Other Trips 1981 %
Car	35	(3)	26	52	(7)	43
Public transport	40	(72)	21	26	(64)	13
Walk	20	(15)	51	16	(11)	41
Other	5	(10)	2	6	(18)	3
	100	(100)	100	100	(100)	100

Source: Great London Council (1985a)

Figures from 1954 are only roughly comparable due to differences in the definition of areas and trips (London Transport Executive, 1956)

are designed to cater first and foremost for the car-borne shopper (see Chapter 7). The modal split of shoppers to such locations is markedly different from the pattern of travel to the established centres. Table 8.6 gives the overall modal split for shopping trips to eight inner London and 21 outer London centres in 1981. Alongside, for comparison, are equivalent figures for Brent Cross, and later figures for two out-of-centre Sainsbury stores.

The London and South East Regional Planning Conference (SERPLAN, 1988, 23–4) has shown that the majority of major retail developments are now in out-of-centre hypermarkets, superstores, retail warehouses and mixed schemes. The SERPLAN report acknowledges that such developments 'are principally designed to serve the shopper using private transport and we have recognized that less mobile shoppers could be disadvantaged in this respect'. This trend in retailing is inevitably reinforced as developers gear their planning and marketing to the new pattern. This is exemplified in promotional literature distributed by the Unit for Retail Planning Information in 1989 which begins, 'As more and more new developments focus on accessibility to the car owning population as a key determinant of demand, the availability of accurate drive time isochrones becomes of

Table 8.6   Mode of travel to shop in London's suburban centres

Mode of travel	Established centres Inner London (8 centres) 1981 %	Outer London (21 centres) 1981 %	Inner London Sainsbury Nine Elms 1985 %	Car-based developments Outer London Brent Cross 1981 %	Sainsbury Hornchurch 1985 %
Car	19	30	57	71	71
Public transport	29	30	14	21	11
Walk	51	38	24	7	17
Other	1	2	5	1	1
	100	100	100	100	100

Source: Greater London Council (1985a) and London Boroughs of Hammersmith & Fulham and Lambeth (1986)

Table 8.7   Travel to three outer London office locations, 1988

| | % Employees | | | | % Visitors | | | |
	Car/ Taxi	Rail	Bus	Walk	Car/ Taxi	Rail	Bus	Walk
Office location A (Uxbridge)	89	3	5	3	84	13	2	1
Office location B (Hayes)	66	1	21	12	97	0	3	0
Office location C (Ruislip)	79	8	7	6	100	0	0	0

Source: London Borough of Hillingdon (1989)

increasing importance'. There is no mention of 'walk times' or 'bus times', nor of the majority of the population who must depend on these alternatives (Unit for Retail Planning Information, 1989).

Shopping, because it is a universal activity, still retains a significant proportion of travel by non-car means at least to older retail centres. The business community, however, has proceeded further down the road of total reliance on the car, and again the extent of this increases with distance from the central area. As an illustration, travel to three office buildings in outer London is given in Table 8.7. Although on the fringe of Greater London, each site is close to a rail station and to a suburban centre which is served by bus routes. They thus represent sites which – for outer London – have above average public transport accessibility. Despite this, the dominance of the car and the marginal role played by public transport and walking is clear. Business developments located away from centres will be even more reliant on the car.

Conclusion

London is a metropolis whose large and dominant central core is dependent upon links to vast suburban areas via high capacity radial railways. During the second half of the 1980s, increases in employment in the centre – particularly in the financial sector – reversed the decline in commuting to the centre. A major issue for London is how best to meet increases in commuting to the centre, as more land is developed in the City of London and Docklands for further office growth.

The prime factors in the deterioration in commuting conditions to central London are the decline in bus use and the reduced rail commuter services. Bus passengers have transferred to rail, where less trains are offered than before, or to car, thus contributing to congestion and the further deterioration in bus service quality.

A new wave of rail investment is possible, following government-commissioned reports on rail possibilities for central London and for Docklands. But restoring rail service frequencies to 1960 levels would in itself bring about a substantial increase in capacity and thus relieve congestion. British Rail have estimated that a 30% increase in peak hour capacity could be achieved 'within five years without major new infrastructure' (Bridge, 1989) which compares with a forecast increase in demand of nearly 20% by the year 2001 (Department of Transport, 1989a, 5). Another possible strategy, promoted by the Association of London Authorities and others, is to strengthen traffic restraint using road user charges (road pricing) in order to reduce congestion and thus allow expansion and improvement of the bus services (Association of London Authorities, 1989).

The continued dominance of London's centre, as an employment area at least, seems assured. Attempts are still being made to increase the capacity of major radial roads, a policy which is criticized for exacerbating rather than helping the commuting problem, but which is unlikely to affect the overall strength of the centre. Rail will continue to perform its specialist role of getting commuters to and from central London, but its share of the overall travel market is small, thus making the system expensive to maintain and develop.

Less certain is the future of transport in London's suburbs. We have discussed the process of motorization in London and argued that the traditional structure of the suburban areas is under threat. Out-of-centre shopping, employment, leisure, health, education and other facilities have reduced the relative importance of the older established suburban centres, and have generated travel patterns that suit the car but not public transport or walking or cycling.

Suburban centres like Harrow, Kingston and Sutton have fought back by pedestrianizing shopping streets, providing more off-street parking and by-pass roads, and improving bus stations. Attempts have also been made by some boroughs to limit out-of-centre developments. The question is to what extent these counter-initiatives are able to check the powerful pressures for motorization? Factors tending to support the maintenance of traditional centres are the existing commercial and other interests there, and the arguments for less reliance on the car for environmental, safety and equality of access reasons. Factors tending to further the progress of motorization are the continuing increases in car ownership, especially the increase in households with two or more cars, the reducing relative importance of work journeys, the shift from the bus, the provision of more road and parking space, and the attraction for developers of out-of-centre sites where parking can be provided relatively easily. All of these factors lower the density of development and make the city less compact. This in turn makes it progressively more difficult to limit the motorization process.

Given its pre-car structure, London's ability to absorb motor vehicles is remarkable. Lack of space has deterred car ownership only at the very centre of the city and a few other high density areas. Restrictions on parking have spread over wide areas of inner London and to most suburban centres, but as a matter of policy there has been little restriction on car *use*. Nevertheless, a heavy price has been paid by London as a whole for the benefits brought to car users. Congestion on the roads has led to extra costs and inefficiencies, and to road speeds no faster than in the 1930s. Bus services have deteriorated to the point where their credibility is in doubt. Environmental degradation has spread throughout the capital, and has reached appalling levels in inner London. The emergence of the car as the dominant mode of travel has also meant shrinking opportunities and choices for those without cars or who depend on others to drive them.

The tensions and conflicts have been experienced since the 1960s in outer London, and have spread increasingly to inner London. Communities on the conurbation fringe are largely accepted as being car-dependent (except of course by those without access to a car) just as driving to work in central London is accepted as being impracticable by the great majority of Londoners. Between these two extremes, the battle between the forces of motorization and the powerful interests vested in London's established economic and physical infrastructure is being fought over the vast swathe of disputed territory which lies between the green belt and the central area. The effect of this dispute to date has been the development of an environmental and transport crisis. A wide range of policies and measures are constantly put forward and debated by a plethora of organizations, but they are often ad-hoc and fail to address the fundamental issues. Almost everybody of informed opinion has recognized the organizational gap left by the abolition of the Greater London Council in 1986, and are united in the belief that some form of single authority is essential if London's transport problems are to be tackled effectively.

Note
1 During proof reading the Government announced a commitment to a crossrail link between Paddington and Liverpool Street.

9

London's Population Trends: Metropolitan Area or Megalopolis?

Anthony M. Warnes

Introduction

London has dominated the settlement hierarchy of   Wales for more
than a thousand years and now accounts for bet   nd 40% of the
population of these two countries. Indeed the first p   dying London's
population geography is to decide how extensive it   this depends all
its other demographic characteristics. If the city is tak   City of London
and the 32 boroughs (the London of this book),   high prevalence
is found of young adults, recent migrants, racial minorities and of districts of
population decline. Arguably, however, the modern city covers a much wider
area, which extends beyond the officially-recognized South East Region. The
'Greater South East Region' encloses peripheral zones with high rates of growth,
relatively high fertility and few overseas-born, and defines a metropolitan region
(or London megalopolis) which has had continuous population growth since the
Second World War.

This chapter reviews selected major themes in London's population geography.
Because London's fortunes have been inextricably linked to those of Great Britain
as a whole, and because the demographic characteristics of so complex and large
a city have been established over a very long period, it opens with a broad historical account and an assessment of London's standing among major world
cities. Thereafter the chapter concentrates upon London's population composition and major demographic characteristics at the beginning of the 1990s. The
principal geographical themes to be covered are those significant for the city's
social change and its management and government. They include the decentralization of London and the formation of a diffuse megalopolis, the migration
relationships between London and the remainder of Great Britain, and its internal
patterns of fertility, mortality and age structure.

Several of London's population characteristics have been present for 500 years
or more: districts with high mortality or fertility, the creation and extension of
suburbs, an over-representation of adolescents and young adults, the presence
of a marginalized and destitute underclass, and the exceptional representation
(compared to the rest of the nation) of overseas migrants and religious, cultural
and ethnic minorities. Such ever-present characteristics, like the 'rise of the middle
class', can confuse if careful distinction is not made between dominant and subor-

dinate features. To take the example of suburbanization, a suburb in Southwark was established in Norman times and they have multiplied ever since, but the mass movement of people into peripheral dormitory districts did not begin until the third quarter of the nineteenth century. The foundation of suburbs should be distinguished from mass suburbanization, which redistributes a large proportion of the total city population. Similarly, while London has for many centuries had a relatively cosmopolitan and ethnically-diverse population, only in the last 40 years has its racial and ethnic composition become a major factor in the city's social conditions. On the other hand, one attribute has declined in importance. Early in this century London lost its former and persistent characteristic of exceptionally low life expectancy.

While some themes in London's demography endure, during the last 150 years the capital's population has been transformed. As a distinct and economically-unified metropolis, London arguably is now declining in population, even if, as shall be examined later, a wider constellation of settlements and labour-market areas in South East England continues to expand. Death rates are unprecedentedly low, are no longer significantly higher than in the remainder of southern England, and are more favourable than those of northern Britain. Birth rates have also fallen substantially, and many associated socio-demographic characteristics of London life have changed: the ages at which children are born, the rising prevalence of divorce and divorced people, the smaller size and changing composition of household groups. And finally, the racial and religious composition of the city has altered greatly. Racially, linguistically and in terms of religious beliefs, until the 1830s London's population was by and large homogeneous: now it is truly cosmopolitan.

The Size and Importance of the Metropolis

'London's importance is a recurring theme in the development of English society and economy. It is therefore difficult to study population trends in England without reference to the experience of London' (Finlay, 1981, xi). The converse also applies, for London's past phases of growth and its present population stagnation have reflected the performance of the United Kingdom economy. Through most of its modern history, London's population has grown more rapidly than that of the nation. In 1100, shortly after the Norman Conquest, London's population was about 15 000. By 1500 it had grown threefold, and in the last 500 years its growth has been two-hundredfold, with particularly strong surges of expansion during, first, the late sixteenth and seventeenth centuries, and then the nineteenth century (Table 9.1). The earlier surge saw London outgrow Amsterdam and Paris (although Naples and Constantinople remained more populous) and was associated with England's emergence as the prime European mercantile power.

During the medieval and mercantile periods, London's share of the national population grew from one-in-forty to nearly one-in-ten, but its importance in England's economy and demography was far greater, particularly in terms of the share of the nation's births that it 'consumed'. Until the middle of the nineteenth century the city's mortality was comparatively high: deaths normally exceeded births, so the city neither sustained itself nor was capable of growth (Wrigley, 1967). In order to expand, London had to import people. For centuries it tempted young migrants with work, prosperity and social mobility, but it gave the majority wretched, insanitary and congested living conditions. Pre-

Table 9.1　Estimates of the population of London, 1500–1901

Year	Estimate (thousands)
1500	50
1600	200
1700	575
1800	900
1851	2363
1901	6586

Source: 1500–1800:　as compiled by Finlay (1981, Table 3.1) from estimates by Wrigley (1967) and others. 1851:　census population of area later administered by the London County Council. 1901:　census population of the area administered from 1965 to 1986 by the Greater London Council.

mature deaths were common, especially among infant children. While this baleful mortality has gone, it remains the case that hundreds of hopeful migrants achieve little or only modest success. The city's haphazard work, high cost of living and encouragement to debt, trap most in the lower social ranks.

From the seventeenth century, London has increasingly dominated the economic and political life of the nation. Around 1800 London and its suburbs became the first 'million-city' and was growing rapidly in association with the country's industrial transformation. Nonetheless, during the early decades of this second surge of growth, the midland and northern cities of England temporarily grew faster than the capital. Only from the second quarter of the nineteenth century did the increasing scale and diversifying fields of economic activity in Great Britain enable London-based bankers and companies to reassert the capital's controlling role. For 50 years, London forged ahead at the apex of the world's settlement hierarchy:　the population of the urban area increasing four-fold from around 1.5 million to around 6 million and its share of the England and Wales population increasing from one-in-eight to one-in-five by 1875 (Table 9.1; Salt, 1986). This was followed during the final quarter of the last century by a phase of relative stagnation, matching the nation's faltering economic dynamism. Britain lost ground in relation to its competitors and London's population lead over New York slipped away around the First World War.

Since 1900 London's position in the rank order of world cities has steadily fallen. Major cities of the United States, Asia and Latin America have successively overtaken it in population size if not in economic strength. London's world position has fallen from first to twelfth (Figure 9.1). Within Europe, the comparatively strong economic performance and the late surge of urbanization in France after 1950 enabled Paris to close the gap on London. London's future economic role and population ranking is highly uncertain. During the 1980s its economy has appeared relatively strong, but widening differentials in job creation and new firm formation between the capital and the rest of the country reflect deep-seated weaknesses in the United Kingdom economy. The imminent prospect of a common European market and of more efficient rail, air and electronic communications with our neighbours in the European Economic Community may herald a shift across the English Channel of economic dynamism and of ambitious migrants.

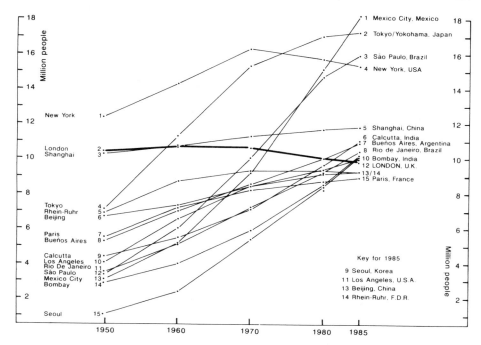

Figure 9.1 Growth and world ranking of the London metropolitan area, 1950–85

Source: United Nations Department of International Economic and Social Affairs data tabulated in World Resources Institute, <u>World Resources 1986</u>. Basic Books, New York, 1986, Table 3.3, p.252.

The Extent and Size of Modern London

Neither the area physically occupied by contiguous buildings nor the distribution of urban land uses are now reliable indicators of the extent of a metropolis. In 1915 Patrick Geddes used the term 'conurbation' in *Cities in Evolution* for an aggregation of interdependent urban areas. He described London's population distribution as sprawling like a 'polypus' (rather than 'octopus') far beyond the bricks-and-mortar of London proper. The difficulty in defining the extent of a modern city arises from three related economic and demographic trends: population decentralization, the similar redistribution of economic activity and jobs – which sometimes leads but normally follows residential decentralization and the decreasing containment of the urban economy within the metropolitan region.

London's economy is increasingly characterized by services-production and intricate and far-ranging spatial patterns of links between producers and customers. Telecommunications, mechanically-driven vehicles and single-purpose high-speed routeways have encouraged the dispersion of both residences and production, and more complex, specialized and inter-dependent systems of production encourage larger urban scale. As in most western cities, therefore, London's population has simultaneously increased and decentralized; residential densities at the centre began to fall from the 1840s (Clark, 1951). The pervasiveness of suburbanization was noted widely a century ago: '. . . the centre of population is shifting from the heart to the limbs . . . the people of London will dwell in urban districts . . . far into the home counties, the clerk and trader will remove to remote suburban villages, as the merchant and stockbroker move further afield to the Sussex Downs and the Hampshire commons' (Low, 1891, 553; Weber, 1899; Wells, 1901).

For much of this century functional connections have been useful indicators of a city's extent, such as the flows of commuters from outlying residential areas to and from central employment areas (Warnes, 1980). Even this procedure is now compromised, as employment (also shopping and administrative) centres have dispersed into a scattered constellation. The transactional and communications contacts that bind a modern metropolitan area are now a loose pattern of cascading links with no clear break between 'within-city' and 'between-city' connections. In the 1960s Jean Gottman employed the term 'megalopolis' to describe the evasive coherence of a modern metropolitan area, with specific reference to the complex along the north-eastern seaboard of the United States dominated by the formerly autonomous cities of Washington, Baltimore, Philadelphia, New York and Boston (Gottman, 1964).

Although London's evolution from a single nucleated settlement has been quite different, it is now a similar fragmented urban region. Croydon, Watford and Dartford are firmly embedded in the physical and economic core of the London megalopolis. Gatwick Airport and Crawley are indisputably integral parts of the London functional region. Today's taxonomic arguments are about the degree of independence of surrounding 'labour market areas', such as those of Chelmsford, Maidstone or Reading, at least 30 miles from the metropolitan centre, and even of more distant Brighton, Basingstoke, Oxford and Southampton (Champion *et al.*, 1987). As the rapid population and economic growth of these areas is related to London's proximity and to their attractiveness for serving the metropolitan market and for 'overspill' housing developments, the city should now be described as a megalopolis similar to Randstad in The Netherlands or to Greater Los Angeles.

Peter Hall's (1989) *London 2001* re-examines the structure and extent of the metropolitan region and notes that the highest rates of housing and population growth have recently moved outwards at a rapid rate. During the 1950s, the maximum growth zone was 15 to 30 miles from the centre; by the 1960s it was 35–70 miles out and, in concentrating on Reading–Basingstoke, Southampton–Portsmouth and Milton Keynes, it had begun to fragment. By the 1970s the fastest growth zones spilt over the boundary of the South East Region. Adopting a suggestion by the Regional Studies Association, Hall argues that the spatial-demographic field of the London metropolis now extends into the 'South East Fringe'. This Fringe, comprising the fast-growing counties of Dorset, Wiltshire, Northamptonshire, Cambridgeshire and Suffolk, surrounds the familiar four-fold zonation of the South East Region: inner London, outer London, the outer metropolitan area, and the Outer South East Region (see Figure 9.2 for the areas of these zones). The five zones together make up a 'Greater South East' region which is here adopted as the megalopolitan region of London in the early 1990s.

The population of the Greater South East increased from 18.2 million in 1961 to an estimated 20.4 million in mid-1988, so that it now makes up 40.5% of the total population for England and Wales (Table 9.2). Since 1961, a marked redistribution within the megalopolitan region has occurred. The share within London itself (i.e. the former Greater London County) has fallen from 44.0% to 33.0%, and in the officially recognized London Metropolitan Area (GLC plus the outer metropolitan area) from 68.1% to 60.1%. The outer metropolitan area's share grew from 1961 to 1981 but has since fallen. It is the outermost regions

Figure 9.2 The London megalopolis

whose share has risen most: the outer South East Region has gained 5.1 points to 24.8%, and the Fringe 3.1 points to 15.1%.

This evolving growth pattern is indicated by the comparative growth rates of the different urban zones in successive periods. During the 1960s the outer metropolitan area's average annual growth rate of 1.72% was considerably higher than all other zones, but by the 1980s its rate was less than one-fifth of those in the outer South East Region and the Fringe. By 1981–88 a monotonic positive gradient of annual rates of change had been established, from –0.03% in inner London, to +1.24% in the Fringe (Table 9.2). There has been an unexpected and notable recovery of population growth during the 1980s: even London's population is increasing, as a result of higher fertility and a lower net out-migration rate (Champion, 1987a, 1987b; Champion and Congdon, 1988).

The most recent and sophisticated description of the spatial structure of the urban regions of Britain by the Centre for Urban and Regional Development Studies employs data from the 1981 population census on journey-to-work flows and population growth rates (Champion *et al.*, 1987). It describes a functional hierarchical urban region with a contiguous '*London Metropolitan Region*' core of over 12 million people extending from Basingstoke to Chelmsford and from Bishop's Stortford to Horsham. Around this lie two 'freestanding' metropolitan areas, Brighton and Portsmouth, which, with the remainder of the South East and beyond, are interconnected in a progressively more intricate web of commuting relationships and other forms of interdependence.

Table 9.2 Population in five zones of the wider London region, 1961–88.

	Population (millions)								Av. Ann. Change (%)		
	1961		1971		1981		1988		1961–	1971–	1981–
	No.	%	No.	%	No.	%	No.	%	1971	1981	1988
Inner London	3.5	19.2	3.0	15.4	2.5	12.8	2.5	12.2	-1.41	-1.41	-0.03
Outer London	4.5	24.7	4.4	22.6	4.2	21.4	4.2	20.8	-0.18	-0.47	0.09
Outer metropolitan area	4.4	24.2	5.2	26.7	5.5	28.0	5.5	27.1	1.72	0.48	0.20
Outer South East	3.6	19.7	4.3	22.1	4.6	23.5	5.1	24.8	1.70	0.79	1.03
South East fringe	2.2	12.0	2.6	13.3	2.8	14.3	3.1	15.1	1.39	1.00	1.24
London	8.0	44.0	7.5	38.5	6.7	34.2	6.7	33.0	-0.70	-1.04	0.05
London metropolitan area	12.4	68.1	12.7	65.1	12.2	62.2	12.3	60.1	0.22	-0.39	0.12
South East	16.0	88.0	16.9	86.7	16.8	85.7	17.3	84.9	0.58	0.05	0.28
Greater South East	18.2	100.0	19.5	100.0	19.6	100.0	20.4	100.0	0.67	0.18	0.42
England & Wales	46.2		48.7		49.1		50.4		0.54	0.08	0.36
Greater South East share		39.4		40.0		40.3		40.5			

Source: Censuses of Great Britain 1961, 1971 and 1981, and OPCS (1990)

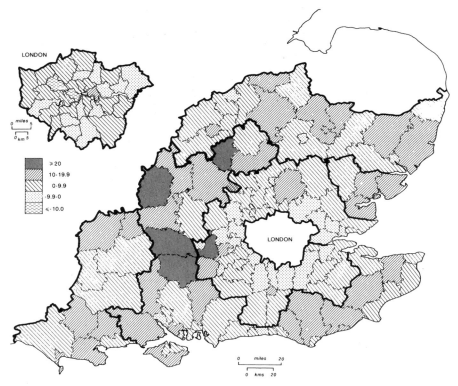

Figure 9.3 Population change 1981–88 by local authorities in the Greater South East

The distribution of population growth during 1981–88 demonstrates that the main areas of growth have moved beyond the outer metropolitan area. The 1988 mid-year estimates of population moderate earlier estimations of growth, as mapped by Hall (1989, 42). Nonetheless, one quarter of the 163 local authority districts in the Greater South East had increases of more than 10%. Only one is in London, Tower Hamlets, where a 25% increase marks both the redevelopment of Docklands and the high housing densities of its recent South Asian immigrant population; and only four are in the outer metropolitan area (East Hertfordshire, Crawley, Bracknell and Wokingham). Only the Brighton–Worthing conurbation to the south of London breaks an annulus of local authority districts with population increases in excess of 10% after 1981 (Figure 9.3). This ring, approximately 45–70 miles from the centre of London, links Canterbury and Ashford in Kent with the East Sussex Weald, the Hampshire and Berkshire Downs, Oxfordshire, Buckinghamshire, Cambridgeshire and the north Essex and Suffolk coasts. Many of the major rail improvements during the early 1980s have improved and electrified train services into this zone (e.g. to Cambridge, Peterborough, Bedford and East Grinstead). The rising commuter flows that are both the cause and the consequence of these service improvements are the visible evidence of inexorable expansion.

On the other hand, the areas of current population decline are virtually confined within the London metropolitan area. The five exceptions are the tightly-bounded towns of Corby, Cambridge, Ipswich, Southampton, Gosport and Shoreham (the Adur District). Within London, 14 of the 32 boroughs had a fall in population during 1981–88, but only in the City of London, Westminster and Brent did the loss exceed 10%. The main areas of population loss are in west-inner London, south-outer London and a contiguous ring of authorities in the outer metropolitan area on all sides except the north-west. With growth predominantly in the outer South East Region and beyond, here is further confirmation that a megalopolitan scale of interacting demographic effects applies to London.

Mechanisms, Processes and Causes of Current Population Change

In analyses of areal population change it is important to distinguish mechanisms from underlying processes. Fertility and mortality determine natural change, while actual change is also affected by net migration gains or losses: all three might be called 'proximate causes'. There are no necessary interactions among these three vital rates of demographic change, and each may be influenced by different social and economic factors which change over time. The task of understanding the underlying causes of areal population change is compounded by the limitations of data sources including their use of artificial time periods and 'convenient' geographical areas. Successive phases of population change do not arrange themselves conveniently by census decades, neither do patterns of loss and gain, or of migration origin and destination, respect political (or geographers') boundaries.

To appreciate the dynamics of population change for an entire megalopolis and its constituent regions, knowledge about national, regional and local trends

Table 9.3 Vital and migration statistics for Greater London, South East England and England and Wales, 1971–87

A   Migration Transfers

Period	Inner London		Outer London		Greater London	
	Natural change	Migration change	Natural change	Migration change	Natural change	Migration change
1971–76	1.0	−20.9	1.5	−7.6	1.3	−13.0
1976–81	1.0	−15.9	1.2	−4.8	1.2	−9.2
1981–86	3.2	−6.2	2.2	−1.8	2.6	−3.5

B   Vital Statistics 1987

	Crude birth rate	Total period fertility	Per cent outside marriage	Crude death rate	Standard mortality ratio	Deaths under 1 year per 1000 live births
Greater London	14.9	1.85	26.5	10.6	95	9.5
Borough mean	14.8	1.84	25.9	10.6	95	9.8
Coefficient of variation (%)	15.0	12.9	34.6	8.1	8.0	31.5
Rest of South East	13.2	–	17.8	10.7	–	8.6
South East England	13.9	1.81	21.5	10.6	94	9.0
England and Wales	13.6	1.81	23.2	11.3	100	9.2

Source: OPCS (1989a; 1989b)

is required in at least four fields: the vital rates, external migration, economic and employment trends, and housing and physical planning developments. The following commentary draws from a selection of statistics on the demographic mechanisms or components of change (Table 9.3) and a chart which identifies some of the interactions between the underlying causes and the population outcomes which have been most important since the early 1970s (Table 9.4). The chart identifies successively the mechanisms by which London's population size and structure are changing, the economic and social factors in London's relative growth, and the factors producing redistribution and internal population differences.

The main points are that mortality differentials are now of little importance in generating either the relative growth of the London megalopolis or areal variations in growth within the region. Mortality improvements over the last 150 years have been associated with an improving standard of living. Rising incomes and improving living conditions have characterized most social and occupational groups throughout the country. Inner London no longer has generally high mor-

Table 9.4 Relations between national, regional and local socio-economic factors and the components of population change in contemporary London

Components of Population Change		
Mortality	*Fertility*	*Migration*
A *'Outcome' Vital Statistics*		
Deaths per 1000	Births per 1000	Net migrants per 1000
B *'Proximate' Demographic Mechanisms*		
Age structure – share in older high mortality and young low mortality groups. No. of births, / infant mortality.	Age-sex structure: ages and married fraction of females of reproductive ages. Rate of first marriages.	In-migration from the rest of the World Out-migration to the rest of the World.
C *Socio-economic Factors Influencing the London Megalopolis*		
Concentration of high-risk groups (drug addicts, homosexuals, mentally-ill homeless) may raise London mortality rates.	Present concentration of very recent immigrants, e.g. Bengalis, may sustain high fertility short term.	Prosperity of London relative to rest of UK affects rate of young adult in-migration. High housing costs deters young migrants and encourages out-migration from mid-20s. Rate of new house completion affects rate of relocation.
D *Socio-economic Factors Influencing a London District or Zone*		
Age-sex structure and over or under-representation of high and low risk groups; elderly share.	Age-sex structure and over or under-representation of women in early 20s and of high-fertility minority groups.	Availability of housing land. Amount of commercial development. Proximity to major growth centres, e.g. Channel Tunnel, Stansted, M4 hi-tech corridor, Milton Keynes.
E *Principal Factors Influencing Long-Term Trends*		
Gradual improvement in rates at all ages likely to continue. Decline of urban –rural differentials may cease.	Fertility rates unstable in modern populations. Rise of illegitimate births to continue. Convergence of birth rates among ethnic groups to continue.	Post-1914 trends for rate of mobility to fall and for mean distance to increase likely to continue. In-migration from non-EEC to remain low, but growth of intra-EEC moves.

tality, but there is some evidence of high death rates in the East End and of a reassertion of pronounced mortality variations amongst recent immigrant groups and multiply-deprived populations (Congdon, 1989).

Temporal and geographical alterations in fertility have recently been more significant. High birth rates generated relatively rapid growth in the United Kingdom and metropolitan London populations during the 1960s, while low birth rates were the principal 'proximate cause' of low population growth during the 1970s. Local fertility differentials are also important, as evinced by the unusually high birth rates in east London boroughs during the 1980s. But the most important mechanism for generating differential growth between London and the rest of the country, and among its own districts and zones, is net migration.

Internal and international migration rates are somewhat easier to comprehend, although no easier to forecast. External migrations are clearly influenced by relative opportunities within Britain and in the rest of the world but in-movements are heavily constrained by immigration law. Internal migrations are closely related to new household formation (itself partly demographically determined by the numbers reaching early adulthood) and economic growth, perhaps more through the level of housing investment and the stimulus to migrations caused by the occupation of new dwellings.

Fertility and mortality

Fertility is slightly higher in London than in the rest of the South East, which in turn has a higher birth rate than England and Wales (Table 9.3). A large part of these differences is explained by age-structure differentials, but there are only modest variations in the 'Total Period Fertility Rate' (the number of children that would be born to a female if present age-specific birth rates prevailed throughout her reproductive years). Within the Greater South East a great diversity of areas have relatively high fertility. The local authorities in the region with 'Total Period Fertility Rates' above 2.00 in 1987 were eight London boroughs, two new towns (Harlow and Stevenage) and four industrial towns (Gravesham, Luton, Slough and Thurrock) in the outer metropolitan area, along with Corby, Eastleigh, Ipswich and Swale in the two outer zones (OPCS, 1989b).

In 1987 the variation in fertility rates among London boroughs was roughly double the variation in mortality rates. Indeed the local authority districts with the highest and the lowest fertility levels in England and Wales were in the London region. Taking first crude birth rates (for the nation 13.6 per 1000), the three inner-east London boroughs of Newham (20.1), Hackney (19.9) and Tower Hamlets (19.2) topped the rank order of districts, while the lowest figure (8.5) was for Rother in East Sussex (covering Hastings and Bexhill). Newham (2.35) also had the highest 'Total Period Fertility Rate' and Camden the lowest (1.31).

A striking characteristic of births in London and the South East is the high median age of mothers: 30.9% of all births in London are to mothers aged less than 25 years, compared to 31.2% in the rest of the South East, and 39.9% in the rest of England and Wales. On the other hand, 34.5% of London births are to mothers aged at least 30 years, compared to 31.7% in the rest of the South East, and 25.5% in the rest of England and Wales. Only a small part of these differentials is accounted for by age structure differences: occupational variations and career or familistic orientations may be stronger influences. There is also a strong geographical pattern within London and the outer metropolitan region in the percentage of all births that occur outside marriage: Lambeth (46.3%) and Southwark (43.6%) have the highest 1987 rates of any districts in England and Wales, whereas the peripheral suburb of Wokingham (8.8%) had the lowest.

In the 1980s mortality in London tended to be higher than in the remainder of South East England but lower than in England and Wales or Great Britain (Table 9.3). The 1987 London Standard Mortality Ratio for deaths at all ages (which applies age-specific death rates in a region to a standard age structure) was 5% lower than the nation's, whereas in South East England as a whole it was 6% lower. However, there is considerable variation within the Greater South East with, in general, areas of highest mortality being in inner London and many of the more affluent rural districts and Home Counties towns having among the most favourable mortality figures in the country. The only local government district in South East England which has mortality comparable to the worst found in the Northern Region of England, where the death rates are 20% higher than the national average, is Tower Hamlets. The other districts in London with relatively high mortality only reach levels close to the middle of the distribution among Northern Region authorities (Townsend *et al.*, 1988, 25).

A recent analysis of mortality during 1981–87 in London found that the 10% of electoral wards with the highest social deprivation scores had an average standard mortality ratio of around 114, whereas the ratio for the least deprived decile was 86 (Congdon, 1989, 484). Overall mortality variations continue to be a cause of concern and are a sensitive indicator of areas and populations that are seriously disadvantaged. This is particularly brought out by the more specialized measures of mortality, such as perinatal and post-natal mortality. In 1987 in London as a whole there were 9.5 deaths at less than one year of age per 1000 live births but the variation among the boroughs was unusually high (Table 9.3). Ignoring the City of London (where there was only one such death), the OPCS (1988 51) reported that the highest rates were in Camden (14.3), Hackney (12.9) and Barking & Dagenham (12.2). Yet, apart from still-births and infant mortality, the scale of mortality variations within the region and the differential with the rest of the country are low: mortality has a secondary role in determining the relative growth and the differential age structure of either London or the Greater South East.

Age and sex structure

The age structure of the Greater South East is essentially similar to that of the nation. As in most north-western European regions, the population pyramid is extremely steep-sided, reflecting the low level of mortality (and the high life expectancy) in much of this century. There is a 'bulge' among those aged 15–29, reflecting the relatively high fertility of the late 1950s and 1960s. According to the population estimates for mid-1987, which publish population data by sex and for eight unevenly bounded age groups, the Greater South East has approximately a 5% over-representation of people under the age of 30 years, with a peak excess of 8% among young men aged 15-29 years. Incidentally, it is noteworthy that unlike national population pyramids during the second and third quarters of this century, no deficits from wartime losses are evident. There are under-representations of all other age groups, generally slight among those of late working age, peaking at a one-fifth deficiency amongst those in their early sixties, and falling to around 10% among the pensionable population. Women aged 75 or more years are slightly over-represented. These differentials are brought about by differences in past and present fertility and mortality and by the patterns of net migration over the last fifty years.

The five zones of the Greater South East have essentially similar age-sex distributions (Figure 9.4). The most distinctive is inner London, partly because it has the fewest people. It has around 15% more adolescents and young adults (aged

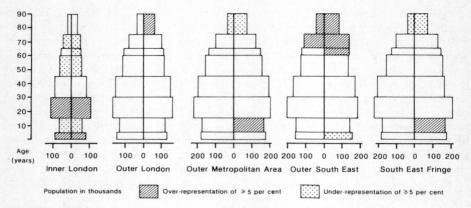

Figure 9.4   Age-sex structure of five zones of the London megalopolis, 1987

15--29 years) than predicted by its share of the megalopolitan population, and pronounced under-representations of late-working age and early-retired people. Reflecting the recent recovery of fertility in inner London, it is also the only zone to have a significant over-representation of children under 5 years (+ 9%). Indeed, the highest single departure from the regional age-sex distribution is the 20% excess of men and women aged 75 or more years in the outer South East Region, which is readily associated with the retirement role of the Essex, Kent, Sussex and Hampshire coastal resorts. At a more detailed geographical scale of analysis, areas of over-representation of elderly people are found in the south-western suburbs of London, the outer metropolitan area and along most of the North Sea and Channel coasts. Exceptional under-representations of elderly people are found in some of the inner London boroughs, such as Islington and Tower Hamlets, and the fastest growth towns of the outer South East Region and the Fringe, such as Basingstoke, Swindon and Milton Keynes (Warnes, 1987, 1989).

Ethnic and racial composition

Chapter 10 of this book provides a detailed account of the history of immigration into London and of its present ethnic and racial composition. Only a brief summary of the most important contemporary facts are given here. In 1981, 18.2% of Greater London's population was born overseas, and about one-quarter of its population lived in households headed by a person born overseas (Table 9.5). The diversity of this population is striking: not only were there substantial communities from the Republic of Ireland, Australia, New Zealand, Canada, the Caribbean, India, Bangladesh, Pakistan, East Africa and the Mediterranean, but several of these groups included people of different races, religions and socio-economic status. It has been estimated from the OPCS Labour Force Survey data that in 1986 there were 958 000 members of the non-white ethnic minorities living in Greater London, about 14% of the total population. The largest groups, with around 300 000 people each, were those of Afro-Caribbean and Indian origin. Only 14% of those of Pakistani origin were in the GLC area, and there were 36 000 of Chinese origin (Table 9.5).

There are strong contrasts in the distributions of ethnic and social groups within Greater London and South East England. People of West Indian origin are heavily

Table 9.5   The ethnic minority population of metropolitan London, 1981–86

Birthplaces of Ethnic Group	Persons in 1981				Persons in 1981 by head of household's birthplace		Self-attributed ethnic group – persons 1986
	London %	Outer Metro. Area %	Outer South East %	South East %	Inner London %	Outer London %	London %
East Africa	1.4	0.3	0.2	0.7	0.9	1.7	0.9*
Caribbean	2.5	0.4	0.2	1.2	8.4	2.6	4.5
India	2.1	0.7	0.5	1.2	2.6	3.9	4.4
Bangladesh	0.3	0.1	0.0	0.2	1.0	0.1	0.7
Mediterranean	1.0	0.2	0.3	0.5	n.a.	n.a.	n.a.
Pakistan	0.5	0.4	0.1	0.4	0.8	0.8	0.8
China	n.a.	n.a.	n.a.	n.a.	n.a.	n.a.	0.5
Mixed/Other	n.a.	n.a.	n.a.	n.a.	n.a.	n.a.	2.3
New Commonwealth and Pakistan	9.5	2.6	1.8	5.1	19.4	11.8	n.a.
Irish Republic	3.0	1.2	0.9	1.8	6.1	3.8	–
Rest of World	5.7	3.0	2.9	3.5	8.9	4.8	–
Total non-UK	18.2	6.8	5.6	11.0	34.4	20.4	–
All ethnic groups	n.a.	n.a.	n.a.	n.a.	n.a.	n.a.	14.2

Source: OPCS (1982) and OPCS (1983), Diamond and Clarke (1989)

Note: The 1986 figures are based on Labour Force Survey data. *All Africa.

concentrated in inner London, with their number dispersed among most wards, except in the northwestern high status belt and riverside residential districts, where either high property prices or discrimination in the private and local authority housing markets has prevented their large-scale entry. There is a western sector in Brent and Hammersmith with its apex at Paddington, a northern segment across Hackney and Haringey, and a broad southern triangle with vertices in Wandsworth, Lewisham and north Croydon (Peach, 1984). The London population of Asian origin has tended to settle in outer London: these boroughs have about half of Greater London's Pakistanis and two-thirds of its population of Indian origin. Harrow has a concentration of Ugandan Asians, whereas Punjabis are concentrated in Hounslow, Ealing, Southall and, in Kent in the outer metropolitan area, in Gravesend. Numerous factors have influenced these contrasting distributions. The rail station or airport of entry has sometimes been influential, the location of employers has been significant and, it has been argued for the concentration of Sikhs in Hounslow, location under the flightpath to Heathrow Airport lowered the price of middle-class inter-war housing and encouraged population turnover (Peach, 1984, 220).

Spatial Population Dynamics in the London Megalopolis

Among the most frequently studied topics in London's population geography are suburbanization, decentralization, peripheral expansion and age-specific patterns of in- and out-migration to the inner and outer areas. All these are variations on a spatial-demographic theme, the emergence through time of an ever more extensive megalopolitan region. The Greater South East continues to grow by attracting to it young and upwardly-mobile people. They tend first to concentrate in the inner area and later many decentralize. As time goes by, their occupational and social status rises, and, as they move, they take, as it were, some of their social upgrading with them. This is a simplified description of a relatively small number of *net* geographical and social movements. They take place against a background of a much larger number of criss-crossing shifts which cancel each other out. But if even only 1% of the population participate in the net changes described above, over the years the city expands, and its social distributions and population geography changes.

London's socio-demographic role in providing education, training and career-launching for a significant minority of the nation's young people is another variant of an enduring theme. Young people are attracted to the Greater South East for further and higher education, for vocational training, and by a buoyant labour market, high wages and high starting salaries. On arrival they tend to concentrate in inner London but when new families are started and secure jobs obtained, many move to the suburbs and satellite towns, and some migrate away. These age-specific in- and out-migration flows condition the age structures of the region and influence demand for primary education, housing and travel, as well as playing an important role in national occupational and social restructuring (Table 9.3). The London megalopolis acts as an academy for training the skilled labour force of the country and for raising the socio-economic status of its young population.

Over a given period the labour force of any region changes its membership and composition. As to membership, there will be a substantial number of exits (retirements) and entries (mainly people ceasing full-time education and taking jobs). Other things being equal, if working ages for men and women today extend from 18 to 62 years, then, making some allowance for mortality and early retirement through sickness, approximately one-quarter of the labour force will exit in one decade, and one-quarter will be new entrants. Although large numbers achieve upward social mobility during their working lives, for some the trajectories are irregular. On re-entering the labour force after child raising many women accept lower status jobs but subsequently achieve significant upward occupational mobility. Many of the compositional changes of the labour force, such as the recent decline in unskilled manual work and the rise of professional and managerial occupations, are therefore brought about by the differential composition of those entering and leaving the labour force.

An early description of the age, sex and occupational composition of migrants into and away from London showed that on average the older out-migrants had a higher occupational profile than the more adolescent in-migrants (Gilje, 1983). The most recent analyses of the 1% of linked records from the 1971 and 1981 censuses have yielded more information on the association between migration into and out of the South East Region and of occupational changes (cf Chapter 6 for a description of this source). Fielding (1989) has shown that among men and women who were in the labour force in both 1971 and 1981, and who over that period moved from the rest of England and Wales into South East England, 30% were in professional and managerial occupations. Another 10% of the in-migrants to the Region entered the labour force in this category, and 24% entered junior white collar occupations (Table 9.6). The high occupational status of continuing workers moving into the region was accompanied by significant occupational up-grading. 26.6% of the migrants improved their occupational ranking, as against 15.1% whose occupational category fell. The net effect therefore was

Table 9.6 Occupational composition and mobility among migrants to and from South East England from the remainder of England and Wales, 1971–81

Occupational group	Migrants from rest of England to the South East			Migrants from the South East to rest of England and Wales		
	Share	Class change 1971–81		Share	Class change 1971–81	
	%	Up (%)	Down (%)	%	Up (%)	Down (%)
Professional, technical and managerial	47.6	36.6	–	40.2	32.2	–
Self-employed and small employers	5.1	57.8	18.3	10.5	51.0	24.9
Lower white collar employees	21.5	21.2	10.9	19.6	16.3	32.8
Blue collar employees	19.8	8.7	25.6	21.2	16.3	32.8
Unemployed	6.0	–	87.6	8.5	–	91.0
All employees	100.0	26.7	15.1	100.0	22.4	20.4
Number of employees	2153	573	325	3389	759	690

Source: Recalculated from Fielding (1989)

Note: The following aggregations of OPCS socio-economic groups are used: Professional, technical and managerial = SEGs 1, 2.2, 3, 4 and 5.1; Small employers and self-employed = SEGs 2.1, 12, 13 and 14; Lower white collar employees = SEGs 5.2, 6 and 7; Blue collar employees = SEGs 8, 9, 10, 11 and 15. Upward and downward mobility indicates a move between these classes, as ranked, from 1971 to 1981, and is expressed as the percentage of migrants in the destination class.

that one-in-ten workers improved their occupational grading in association with migration into South East England.

Out-migration from the South East to the rest of England and Wales by continuing workers was also accompanied by upward social mobility. While it is undoubtedly true that in aggregate migrants are likely to exhibit rising occupational status in association with their greater ages, experience and the fact that promotions stimulate moves, the effects in South East England are pronounced. Out-migration was dominated by the professional, technical and managerial workers but to a lesser extent than in-migration, and with upgrading from the working class to both professional and managerial workers and the self-employed. At the same time, however, out-migration involved the transfer of many men and women into unemployment.

Overall Fielding's (1989) analysis of the OPCS Longitudinal Survey data has provided a more detailed picture of the way in which South East England attracts large numbers of young working-age people from the rest of the country, enables a significant minority of them to raise their occupational status, and simultaneously is a net exporter of workers who in the process consolidate and, particularly for women, further enhance their occupational rise. As he says, the South East '. . . is a region of upward social mobility, a kind of escalator for those who want to "get on" in life' (Fielding, 1989, 35). A noticeable number of professional and managerial workers who move out of the South East, switch into self-employment and the ownership of small businesses: hence 'the region exports its entrepreneurial culture'.

Hamnett (1976, 1986) has compared male occupational changes during 1961–81 in London, the rest of the South East, and England and Wales. He used groupings of the 17 official social economic groups which differ in detail from

Fielding's, but comparisons are possible. During 1961–71 London's loss of 343 000 or 13.0% of its economically active male population was balanced by a gain of 330 000 or 13.0% in the rest of the South East. In the following decade, however, London lost 316 820 or 13.8% while the rest of the South East gained only 56 320 or 2%. London had 51% of the South East Region's economically active males in 1961, but only 40.5% in 1981. In 1961 the socio-economic composition of London and rest of the South East were virtually identical: London had a marginally lower representatiaon of professional, managerial and intermediate non-manual workers (21.0% as against 22.2%), a slightly higher percentage of unskilled workers (8.4% to 6.9%) and more semi-skilled, personal service and junior non-manual workers (32.7% to 29.6%) (Table 9.7). By 1981 the picture had changed significantly. Although both London and the rest of the South East had gains in the professional group of non-manual workers, the numbers and the proportions involved were very different. The increase in the rest of the South East was seven times greater than in London and a gap of four percentage points had been established (31% to 35%). In 1961, the socio-economic composition of inner London, outer London and the rest of the South East was broadly similar in all but two respects. Inner London had a substantially lower proportion of the professional group of workers than outer London (17.0% and 23.8%), and a higher proportion of unskilled workers (11.2% and 6.5%). Both inner London and outer London lost substantial numbers of skilled, semi-skilled and unskilled manual workers during 1961–81, but inner London lost more strongly. Both zones had gains in the professional group of workers during 1961–71, but gains in the outer area were far more than the inner, and in the following decade outer London gained a further 4.5% while inner London lost 4.8%.

Although both London and the rest of the South East experienced marked increases in the proportion of the professional group of workers in their economically-active male population, the increase was not as great as in England and Wales (Table 9.7). On the other hand, the increase of self-employed workers and the decrease of unskilled male employees was greater in South East England than in the entire nation. The area outside London took 88% of the increase in the South East's professional group of workers and by 1981 35% of its male employees were in this category. When these changes are combined with the differential losses in other categories, it can be seen that London experienced a small but significant increase in its proportion of the region's lower status occupational groups during 1961–81. Almost all the absolute increase in the professional group of males in London went to outer London.

Hamnett and Randolph (1984) have provided a valuable analysis of the role of changes in housing ownership and supply in different parts of London and changes in the socio-economic structure of its various boroughs. The gentrification of inner London has been localized, and there have been considerable losses of the professional group of males in the traditionally high status central boroughs. A distinction is apparent between the population and socio-economic changes of the three relatively high status boroughs of Camden, Kensington and Chelsea, and Westminster and the remainder of inner London. During the 1970s, for example, although the whole of inner London had a falling number in the professional group of males, the losses were overwhelmingly concentrated in these three boroughs. In Kensington and Chelsea and Westminster, the losses exceeded 25%. It is only in the more peripheral lower status inner London boroughs, such as Islington and Hammersmith, that increases in the level of owner occupation have been accompanied by absolute increases in the professional group of males. There were small losses in Lambeth and Southwark, but all other inner boroughs

Table 9.7 The occupational structure of economically active males in Greater London, the South East and England and Wales, 1961–81.

Occupation group	Occupational composition 1981			Ratio to England and Wales		Change 1961–81 (%)		
	London	Rest of South East	England & Wales	London	Rest of South East	London	Rest of South East	England & Wales
Employers, managers, intermediate and professional	31.2	35.0	28.0	1.11	1.25	11.0	82.2	48.2
Non-professional self-employed and farmers	6.5	6.8	6.2	1.05	1.10	30.2	76.3	36.8
Skilled manual foremen and supervisors	23.7	25.1	29.6	0.80	0.85	−43.2	−7.1	−18.5
Semi-skilled, personal service, junior non-manual	27.7	23.8	25.4	1.09	0.94	−36.8	−7.4	−20.0
Unskilled manual	5.7	4.2	5.7	1.00	0.74	−48.5	−30.0	−34.6
Inadequately described and armed services	5.2	5.0	5.0	1.04	1.00	46.9	3.2	30.4

Source: Hamnett (1986)

Notes: The following aggregations of OPCS socio-economic groups are used: Employers, managers, intermediate and professional = SEGs 1, 2, 3, 4, 5 and 13; Non-professional self-employed and farmers = SEGs 12 and 14; Skilled manual, foremen and supervisors = SEGs 8 and 9; Semi-skilled personal service and junior non-manual = SEGs 6, 7, 10 and 15; Unskilled = SEG 11; Inadequately described and armed services = SEGs 16 and 17.

experienced significant increases, from 7% in Hammersmith to 29% in Tower Hamlets.

The distinctive experience of these three inner London boroughs is associated with changes in the tenure structure of their housing. In 1961 only 17% of their households were owner occupiers and a further 19% were council tenants. Almost two-thirds of all households rented privately and levels of multi-occupation and sharing were substantial. By 1981 the proportion of households in the private rented sector had been more than halved to 30% while the proportion of owner occupiers and council tenants had increased sharply to 27.0% and 43%, respectively. These changes were far from evenly distributed within inner London: while the council sector grew most rapidly in the eastern, northern and southern boroughs, the increases in the level of owner occupation tended – excepting Tower Hamlets where it had been negligible – to be greatest in those boroughs which experienced the greatest losses of privately-rented households. Thus, both Camden and Kensington and Chelsea experienced increases of 43% in the number of owner occupiers, and Westminster had a 63% increase. The large increases in owner occupation in inner London have primarily stemmed from the sale of previously private rented property. New building for owner occupation has been very limited. These sales have included the 'break-up' and sale of blocks of privately-rented, purpose built flats; the sale of multi-occupied privately-rented property for either single family use or for conversion into flats for sale. All these processes have considerably reduced the number of multi-occupied privately-rented flats (Hamnett and Randolph, 1984). These valuable details of the interaction between housing, occupational change and population change are another illustration of the web of interactions that bind different areas of a metropolitan area or megalopolis together.

Conclusion

Is London's population falling or increasing, and whichever is the case, will that trend continue? Even these fundamental questions are by no means easy to resolve. Although estimates of government demographers for mid-1988 show a revival of population growth in London and, more strongly, in the surrounding outer metropolitan area and outer South East areas, inter-censal estimates have often been wrong. The problems of gaining accurate population estimates, whether through census questionnaires, the registration of births and deaths, National Health Service records, or local authority sources, are considerable. Age-structure effects, notably the presence of the high birth cohorts of the 1960s in the peak child-bearing years, and the concentration in inner London of recent migrants with unusually high fertility, explain the revival of population growth. But this may be a temporary and minor interruption to the lowering of population densities in the inner metropolitan area. By the late-1990s the age groups with highest fertility will be the low birth cohorts of the 1970s. What is more, overseas immigration is likely to be increasingly restricted and the observed trend for the fertility of ethnic minorities to approach the national average will probably continue.

This appraisal of population trends in South East England shows that whether London's population grew or fell during the 1970s depends on its geographical definition. How far does the inner area extend? How extensive is the metro-politan area? Are the areas of current rapid growth in Hampshire, Oxfordshire, north Buckinghamshire, Suffolk and elsewhere signs of the continuing vigour of

London? Few have yet accepted the notion of a London megalopolis, extending 75 miles from the centre, but reference to the long-term patterns of metropolitan deconcentration recommends this unit of analysis. The Greater South East continues to grow faster than Great Britain. Its peripheral zone of fastest growth receives large numbers of young families from areas nearer the centre of London. Its population prospects are dependent upon the success of the London and British economies, not least in a single European market. It will also be strongly affected by the availability and relative cost of housing in the region. Growth may be inhibited by the region's internal inefficiencies and high costs, but there is no sign yet that other regions of Great Britain will restrain the absolute and relative growth of the Greater South East.

10

Race and Ethnicity in London

Emrys Jones

A persistent characteristic of the metropolitan city throughout history has been a marked degree of racial and ethnic heterogeneity. Until recently, great cities have depended for their growth on attracting population from elsewhere and it is not surprising that mere population mobility should have provided a mix of peoples in those cities which most attracted all manner and kinds of migrants.

In London, the extent of this heterogeneity throughout most of its history was surprisingly insignificant. Most 'foreign' groups were small, often with highly specialized skills and functions which were an important adjunct to the working of the city rather than a threat to its well-being. For example, the Lombards established banking in the twelfth century and Huguenot silk weavers came from France along with skilled instrument makers from Holland in the sixteenth century. The late seventeenth century fashion for black servants brought about 18 000 West Africans to London, although they were not a separate community and played no part in the social differentiation of the city. All these, together with continental merchants, diplomats and agents from other countries, were a necessary element of a thriving world city.

The nineteenth century brought a different category of migrant whose numbers were bound to create reactions and pose problems. Although London contained a sizeable Irish population before 1800, following the potato famine of the 1840s Irish peasants flooded into the capital. They were unrelievedly destitute and their language and religion were serious barriers to their easy acceptance. Immediately stereotyped, they became the focus of social prejudice because they were so 'visible'. By the time of this Irish inflow, there was already a considerable number of Jews settled in the City of London and Whitechapel. For most of these settlers, their poverty matched that of the East End, the area which had absorbed them. As with the Irish, differences in both language and religion made acceptance by the host population difficult to achieve. For both these immigrant groups what was established was the association of newcomers with certain geographical parts of the city; the Irish in St Giles and the Jews in the east, producing human ecological patterns which would be strengthened and entrenched with subsequent growth.

Both these migrant groups were an integral part of the population explosion of a great city. But they created neither a pluralist city nor the extreme ghettos of cities elsewhere. Although newer entrants from Ireland retained visibility, older generations lost theirs. Even the more ethnically distinctive Jewish population became partly absorbed as economic improvement and the shedding of orthodoxy made movement and acceptance within the city fairly easy. On the whole, London in the twentieth century settled for a general impression of an overwhelmingly white, Anglo-Saxon, protestant, English-speaking and culturally homogeneous society. In the inter-war period it even escaped the establishment of near-ghettos of coloured seafarers which happened in Cardiff and Liverpool.

In the second half of the century this image of homogeneity changed rapidly and radically. The war itself brought changes, though in such special circumstances that any seeming disruptions seemed acceptable. Such was the case for the large refugee groups, for example. The largest of these was the Polish, which was over 100 000 strong, and in itself was socially homogeneous and closely knit because it was largely drawn from the army officer class and from the professions. Indeed, their initial concentration in the West End amply demonstrated their economic and social status (Patterson, 1964). This, together with their comparative invisibility, made them easily assimilated. Significant changes came in the 1960s with a dramatic increase of West Indian migrants; the visibility of which could not be ignored. Here was another 'race' with features that were unchangeable which offered an obvious peg on which to hang prejudices. London encountered a racial minority 'problem' which was less easily accommodated within the spatial structure of the metropolis than previous inflows, one aspect of which was that new ecological patterns in the distribution of the immigrant population began to emerge (Glass, 1960).

Racism, Race and Ethnicity since 1945

To equate a whole set of cultural and social characteristics with colour may be simplistic, but in essence this has been the attitude of the majority of one people towards another and this forms the basis of racial prejudice. However much academics discuss the idea of 'race' as being biologically untenable, we cannot dismiss attitudes which arise from its popular interpretation. Perhaps it would be better to turn to a sociological definition of race, as a nexus of socially created prejudices which are attached to certain physical characteristics. The degree of 'strangeness' between different groups (and basically the world is divided into 'us' and 'them') increases with colour, and its translation into attitudes and behaviour leads to racism. If the numbers are very small such strangeness appears to be more easily accommodated; if they are seen in any way as a threat then it leads to problems between migrants and the host society.

West Indian migration to Britain began in significant numbers in the 1950s (Peach, 1968). Migrants had been arriving at a rate of about 1 000 a year, but from 1956 to 1959 they averaged 20 000. The majority came to London, so that by 1961 there were 78 000 in the capital, along with 64 000 Asians and 17 000 Africans. At this time annual fluctuations in immigrant numbers were closely linked to the availability of work in Britain. However, in 1960 there was a massive increase in the West Indian inflow to 45 000, partly because of constraints on Jamaican immigration into the United States. This surge in numbers produced an immediate reaction in Britain in the form of legislation which introduced work permits and required that they be renewed every six months. In that year about 100 000 out of a national total of 174 000 West Indians lived in London and the South East, where they accounted for just under 1% of the population. From this point on, West Indian immigration declined somewhat whilst that of Asians – mostly Indians and Pakistanis – increased rapidly. By the early 1960s a quarter of a million Asians had come to Britain and half of these were in London.

Between 1961 and 1971 coloured migrants to the capital doubled. During the earlier period of increase, economists, sociologists and social workers still saw the influx in purely economic terms; not as a 'racial' issue but as a phenomenon of immigration (Peach, 1968). In structural terms the newcomers were seen as a 're-placement' population which was filling specific niches in the economic structure; and had they not been available there would have been an acute labour shortage in some sectors of the economy. Most certainly, there was a time when public transport in London might have come to a halt were it not for migrant labour. It was commonly thought that following economic integration a process of social assimilation would see the absorption of migrants into the general London popu-lation. Indeed, by the 1960s, while it was possible to identify areas where large clusters of West Indians, Asian and African migrants existed, nowhere were they concentrated enough even to hint at the formation of a ghetto (Baboolal, 1981). Thus, although the West Indian population increased by 73% between 1961 and 1971, only a very small number of wards in London had more than 8% of their population in this racial group. Furthermore, between 1961 and 1971, in spite of increasing numbers, there was no trend towards further concentration (Shepherd *et al.*, 1974). Indeed, the percentage of West Indians in 'boroughs of concentra-tion' – those that had an above average number of migrants – actually decreased, and comparable figures for the Asian population were almost static.

Of course statistics mask the social or perceived nature of the situation. People's impressions of reality were quite different from these figures, and the presence of a 'racial element' was enough to identify certain neighbourhoods as being 'West Indian' or 'Asian'. After 1971, the degree of concentration did increase markedly. Yet, even by 1981, there was no indication of ghetto formation. For example, Brent (with 33.5% of its population as coloured immigrants) and Ealing (with 25.4%) had the highest borough proportions, but in Brent only four wards had more than 50% of their population as coloured immigrants. Higher concentra-tions were recorded in Ealing, with one ward scoring 71.0 and another 85.4%, but these are exceptional. Although statistical analyses often refer to segregation, and although all manner of indices have been devised to measure and compare concentrations, it is true to say that there is little or no physical segregation in the strict sense. Scale, of course, has much to do with the measurement of segregation, as well as with its perception; for the smaller the geographical scale of one's focus the more likely one is to be aware of concentrations. But, even at the ward level, a coloured majority is comparatively rare. On the other hand, people's image of social distinctiveness often concentrates on very small areas; horizons often being limited to a street or to a few blocks. This, coupled with the inevitable distortions of personal perceptions, often creates a mental image of intense clustering; which, most especially when accompanied by politically motivated scare campaigns mounted in the popular press and by some politicians (Hall *et al.*, 1978), can invoke fears of threat amongst the host population.

The 1981 census for Greater London showed that just over 18% of its popu-lation was foreign-born; but one-third of these were born in Europe, compared with just one-quarter in Asia and only one-in-seven in the West Indies (with the same proportion from Africa; Table 10.1). There are two important caveats to this simple arithmetical statement. The first is critical to gauging the extent of the coloured population; it is simply that the census registers only the country of birth. The figures are for those born outside Britain. Not only is it possible that these figures possibly underestimate actual immigrant numbers (e.g. illegal migrants who avoid making a census return), but it is also the case that the second and third generation of families who were first generation immigrants are

Table 10.1   Country of origin for London's main population groups, 1981

Country	Number
England	5 110 752
Scotland	108 453
Wales	77 928
Ireland	196 968
Other EEC	90 115
Old Commonwealth	36 512
New Commonwealth	587 896
India	137 424
Pakistan	35 183
Bangladesh	21 817
Caribbean	165 389
East Africa	90 456
Rest of Africa	51 819
Total	6 525 060

not enumerated. Hence, the full extent of the coloured population is not known. Perhaps it is most unwise to guess at what the real numbers are, but it might well mean higher figures by a factor of three or four.

The second caveat is less obvious to the host population but it is very significant to immigrants themselves. It is simply that our census groupings are misleadingly homogeneous, and that, culturally, differences in 'homeland' categories are more significant than their similarities. This has been emphasized by Peach (1984), who contrasts the perceived sameness of, for example, West Indians, with the intense consciousness they have of their islands of origin in the West Indies:

> There is an archipelago of Windward and Leeward Islanders north of the Thames; Dominicans and Santa Lucians have their core area in Paddington and Notting Hill; Grenadians are found in the west, in Hammersmith and Ealing; Monsterratrians are concentrated around Stoke Newington, Hackney and Finsbury Park; Antiguans spill over to the east in Hackney, Waltham Forest and Newham; south of the river is Jamaica. (Peach, 1984, 224)

This is a salutary reminder that dots on the map, or cross-hatching on diagrams, represent people, and although we are forced to make generalizations, at a broad level identities are much more finely gauged, one consequence of which is that patterns of integration, cohesion and confrontation are complex. There are worlds within worlds, and the investigator must participate in the perception of all groups before an understanding of the complex structure of a heterogeneous urban society can be gained.

The Geography of Race and Ethnicity

In analyzing the spatial distribution of racial and ethnic groups in London the underestimation of numbers must always be kept in mind. Hence, all of the maps presented in this chapter only describe the distribution of those born outside Britain (Congdon, 1983). Each of these maps are based on data at the electoral ward level, but as there are 734 wards in Greater London, these data give a detailed picture in spite of not showing some important, more localized,

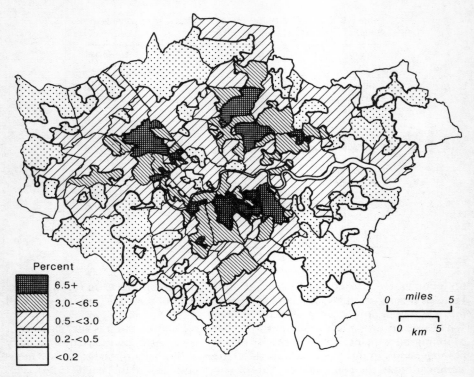

Percent

- 6.5+
- 3.0-<6.5
- 0.5-<3.0
- 0.2-<0.5
- <0.2

0   miles   5

0   km   5

Figure 10.1   Percentage of persons born in the New Commonwealth Caribbean, 1981

variations. We can begin a brief survey of major groups by looking at the West Indian population (Figure 10.1), which with 165 389 persons, is the largest component of the New Commonwealth peoples (accounting for just over 2.5% of London's population). Their distribution is clearly an inner city one. This inner city comprises 12 of the 32 boroughs of the former GLC and is essentially the Victorian-Edwardian city as opposed to the outer ring of inter-war suburban boroughs. Within the most inner core there are very few West Indians; for the broad belt of West Indian settlement is in older, once-decaying, London suburbs (which now comprise part of inner London). Within this older suburban belt there are areas of relative concentration, some of which, as Peach (1968) pointed out, are characterized by specific island groups. South of the River Thames three such nuclei exist, in north Lewisham, in north Southwark and in Lambeth. Some of the earliest post-war migrants came to the Brixton area in the 1950s. North of the river there are two concentrations, one in south Brent, where the foci are Paddington and Notting Hill, the other in east Haringey and Hackney.

The distribution of Asians is rather different (Figure 10.2). It is much wider, more dispersed and with a considerable overlap of Asian subdivisions in the outer boroughs. Although concentrations are less intense than among West Indians, today there are two conspicuous areas which have been well-established for over a decade. The first is in west London, particularly in south Ealing and in Hounslow, the second in Tower Hamlets to the east of the City of London. The point of interest about the first of these is its proximity to Heathrow, which was the point of entry for most Asian immigrants; the second, which is

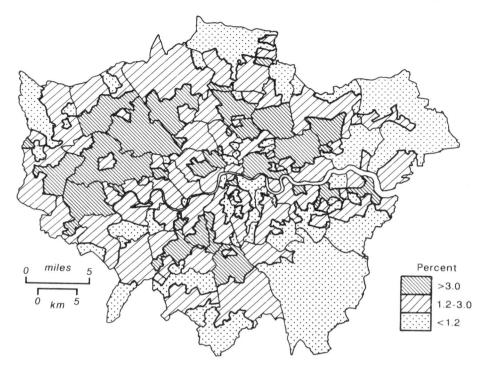

0 miles 5

0 km 5

Percent

>3.0

1.2-3.0

<1.2

Figure 10.2   Percentage of persons born in New Commonwealth Asian countries or Pakistan, 1981

smaller and more concentrated, that it is mainly a Bangladeshi population. This is a reminder that for any interpretive work we should distinguish between the 137 424 Indian, the 35 183 Pakistani and the 21 817 Bangladeshi residents in London (in 1981). There are other areas of relative strength, but they are of lesser importance as population concentrations. Nevertheless, what distinguishes these smaller clusters is that they tend not to be in the inner city and some are very obviously in the outer suburbs, as in central Croydon, Barking, Newham, Redbridge, and the southern part of Waltham Forest.

In the 1960s, there were few East Africans in London, but their numbers were augmented considerably by expulsions from Kenya in 1967 and from Uganda in 1972, the combined effect of which was to add 27 000 to their number. Today, first generation East African migrants comprise 90 000 persons, the vast bulk of which, with the exception of small concentrations in Brent and Harrow, are dispersed across the capital (Figure 10.3). Mostly, these migrants were from the entrepreneurial classes, which, for those who came to reside in the suburbs, accorded well with the middle class lifestyles and aspirations of the host white population.

Moving from racial to ethnic groups, the Irish have had a significant numerical presence in the metropolis for over 200 years (Jackson, 1964; Chance, 1987). Irish-born London residents numbered 196 968 in 1981, and they are strongly associated with their traditional hearth in the north-east sector of the inner city (Figure 10.4). Two direct transport routes from Ireland, those via Holyhead and Fishguard, as well as an indirect route via Liverpool, focused Irish immigrants toward two points of entry in London – Paddington station and Euston station. It

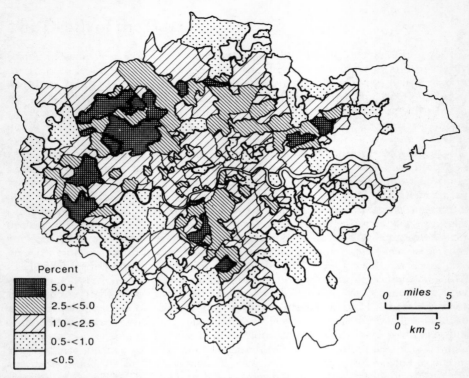

Percent

■ 5.0+

▨ 2.5-<5.0

▧ 1.0-<2.5

⠿ 0.5-<1.0

☐ <0.5

Figure 10.3 Percentage of persons born in New Commonwealth African countries, 1981

is not entirely a myth that the Irish settled within walking distance of one of these stations. With south Brent as the primary concentration, significant Irish numbers can be found in an arc extending from Ealing through north Hammersmith and onto Camden and Islington. Comparatively speaking the Irish-born have a slight presence in the outer suburbs.

Some generalizations can now be made about the groups so far discussed. Again a note of caution must be sounded because it is easy to over-emphasize concentrations. Total numbers are less than coloured groups in New York, for example, and by international standards the degree of geographical segregation is relatively small. Having stated that, it is well to remind ourselves that we are dealing with foreign-born London residents only. There are no data to indicate whether the presence of a second or even a third generation, which for racial groups would mean a retention of racial characteristics, confirms or distorts the distribution patterns that emerge from census maps.

The word concentration has been used in describing the distributions above, but no attempt has been made so far to measure this. So much is perceptual in the social assessment of social mix that doubts may be cast on the usefulness of many of the indices of segregation which dominated this field of academic investigation a generation ago. On the other hand, some objective measurement of social mix is useful if only for comparative purposes. One of the most familiar measurements is the index of dissimilarity, which measures the proportion of any group which would have to move to another ward to give a uniform distribution of an ethnic/racial group throughout the city. Using this measure for London, the

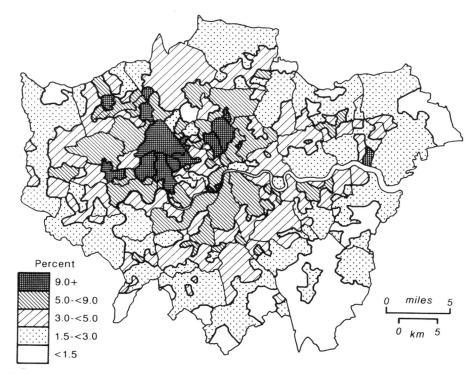

Figure 10.4 Percentage of persons born in Ireland, 1981

lowest value for 1981 is for the Welsh for whom the index score is 16.83; which is lower than that for the native-born English (at 28.52), and provides a clear indication of the assimilation of this group into the host society (Table 10.2). The Irish are more concentrated, but not to the same degree as the coloured population. At 59.42 the dissimilarity index for Bangladeshis is very high. Accepting that this index is a simple statistical measure, it nonetheless shows that there is great variation in the degree of mixing of ethnic and racial groups, and thereby provides some indication of their acceptance by the host population, which is suggestive of the different degrees to which they are integral to the mainstream of London life.

Table 10.2 Dissimilarity index for selected ethnic and racial groups, 1981

Population group	Index
Welsh	16.83
Irish	24.99
English	28.52
Pakistani	45.81
West Indian	46.21
Bangladeshi	59.42

Note: These figures were computed by Dr JM Bland of the Department of Clinical Epidemiology and Social Medicine, St George's Hospital Medical School, who has my thanks for his permission to reproduce the figures shown here.

Accounting for Geographical Patterns

Moving from the description of spatial distributions to their explanation, it has to be stated that the visibility of the Irish has diminished so much in the last century that they can now be considered as more or less 'invisible' (despite the fact that remnants of ethnic prejudice persist, which is periodically given prominence by events in Northern Ireland). While their geographical hearth is firmly established, and most probably has been strengthened by intakes from subsequent generations, their assimilation into society has meant that a large proportion of the population of Irish origin is now indivisibly merged with the host population. That to this day there are many Irish-born in London no doubt means that the traditional links which have existed between the homeland and this metropolis, such as supplying semi-skilled and unskilled labour, still have a role to play in the economy of the capital. But this is not the only Irish contribution, for the medical profession, both doctors and nurses, as an example of just one skilled profession, also relies heavily on recruitment from Ireland.

In contrast with those of Irish background, the coloured populations are more visibly distinctive, as well as having much in common in their geographical distributions. The main impression for such immigrant groups is of concentration in the inner city, and it is very tempting to involve the classic ecological models of American sociologists to explain this: indeed, many of the factors which operate are familiar in this context. Long established changes in the historic core, the financial centre and the West End, means that there are very few inhabitants there. Immediately beyond this, in the inner city, is a zone of ageing buildings which, when not protected by wealth and specialized function, have been left by the upwardly mobile and which is now characterized by high densities of multiple occupation in houses of limited facilities and conditions of degradation. Some of this is the classic initial hearth of the poor migrant.

This simple model merits further examination before it can pretend to have an explanatory role. For example, the inception of coloured migrants within this zone was closely linked with what was happening in London and the South East in the 1950s and 1960s and the housing policies then in force. It was a period of new town expansion when this was seen as the best way of clearing the city of its poorer housing. Whole neighbourhoods were decanted from overcrowded and decaying inner areas to new locations, even beyond the green belt, prior to the rebuilding of central sites at lower densities and with better amenities. The population that moved out was exclusively white, which allowed for the 'colonization' of these inner areas by coloured migrants. In the 1970s the decanting policy lapsed, and housing improvement began *in situ*, thus providing more likelihood that existing neighbourhood configurations were preserved. This in turn led to greater concentrations of migrants in localities that were not dealt with in this way. Their housing opportunities were further restricted by a policy of encouraging local authority tenants to buy their own houses, and these tended to be in the 'better' areas of the city.

Among immigrants the tendency was to occupy privately rented property first (a trend that was encouraged by recent arrivals not having sufficient 'points' on the residency/social need criteria that local councils used to allocate their housing), but by the 1970s many had become buyers and owners − albeit of obsolescent and inadequate property − because this was the primary means of finding accommodation, even if at times this turned out to be overcrowded. By the 1980s two out of five West Indians owned their home, as did seven in ten Asians. Only after some time in Britain could immigrants enter the public sector housing sector. In general, access to housing was beset with constraints, all of which tended to increase the tendency towards concentration. Even opportunities for improved housing often did not lead to a move from what were becoming established areas of immigrant intake. In such conditions desegregation was extremely unlikely. Sometimes a new area opened up unexpectedly. The concentration of Asians near Heathrow is not only a reminder of their point of entry, but was made possible by the unusual availability of houses, and certainly at reduced cost, left by white people who could no longer tolerate the noise of the airport.

Yet the effects on distribution of various mechanisms in the housing market are very complicated, for the allocation of public authority houses and the marketing of the private property are both open to control and manipulation which can result in discrimination and eventually in segregation. Evidence suggests that before the Race Discrimination Act of 1968 house agents exercised considerable discrimination in preserving the non-coloured character of some areas and in directing coloured people to poorer areas. Yet two recent studies by students at King's College London on the outer suburbs found that although there was a tendency for some estate agents to steer clients away from what they thought were unsuitable areas (white from black and black from white) there was not an appreciable difference between what was offered to clients of different colour (Akers, 1987; Sekyi, 1987). Inevitably, there are differences in the individual responses of agents, but it seems reasonable to suppose that there is now little evidence of an institutionalized response.

Similarly, while local authority houses are allocated on a variety of criteria, there is little doubt that over the last three decades or so, with the best will in the world, the cumulative effect has been to some extent discriminatory. A study of Hackney showed that council estates which were dominated by white people when new (and when the proportion of the London population that was coloured was much smaller) eventually tended to become coloured; simply because obsolescence and the emergence of design problems made it difficult to sell or rent to a white population (Commission for Racial Equality, 1984). Offering further insight on racial bias in council house allocation, Phillips's (1987) study of GLC housing in Tower Hamlets found that while overt discrimination was absent amongst housing officers, the fact that they were largely untouched by the need to be actively anti-racist, along with racial harassment by whites on the more attractive estates, did help maintain patterns of geographical segregation.

While discrimination has not been eliminated, therefore, in both private and public sectors its more overt manifestations are less evident today than in the past. But even the least bias, aggravated by economic forces, seems to lead to a sorting process in which colour surfaces as a persistent element. The full explanation of migrant distribution is therefore complex. Even the port-of-entry explanation is a very generalized view of a social process which embodies any recent immigrants' desire to be near their own kind. Once a neighbourhood has acquired some kind of sympathetic association, perception of its migrant status tends to lead immigrants to that area and in such a way that can lead to the formation of 'urban villagers' (Abu Lughod, 1969). Earlier in this century the presence of a Welsh church in Ealing tended to make Ealing 'Welsh' to migrants from Wales who could not be expected to weigh all the pros and cons of every part of a city of seven million (Jones, 1985). The need for community, however loosely defined or indeterminate, and sometimes no more than the need for interaction with people of similar origins, leads to the intensification of such associations and to a greater degree of concentration. Peach (1968) has reminded us of the triggering of distinctive communities by the identification of certain neighbourhoods with specific West Indian islands. The need for community leads to the intensification of association and encourages segregation.

The distributions described here are those of the 1981 census, but there are few radical differences in the pattern from the previous census, or even from that of 1961. It is, however, impossible to judge the totality of the dynamics of the population simply because we lack the data for British-born coloured people. Shepherd and associates (1974) saw little evidence of increasing concentration in 1971, but they were disregarding second generation immigrants. One can safely surmise that there has been some movement to the outer suburbs, otherwise increasing concentration would have been even more marked. We can also say that continuing migration has confirmed the initial pattern because there has been no great change either in the destination of the immigrants or in the inner city conditions which they have met on arrival. There is the expected occurrence between the distribution of immigrants and patterns of deprivation (as indicated by measures like high unemployment) which at an intra-city level we associate with the ageing and obsolescence of the inner city environment. A much more useful analysis may be possible after the 1991 census which will include a question on the racial grouping of those born in this country. Only then will we be able to discuss the full dynamics of the situation.

So far we have assumed a simple equation between residential distribution and race, tending to ignore other considerations. One of the most interesting conclusions drawn by Lee (1977) in his 1970s study was that the West Indian population occupied the same areas of the city as they would if their distribution had been entirely the result of their socio-economic classification. Take away colour and they would still be in inner city sectors of urban decay and deprivation; hence, a socio-economic explanation of the distribution of London's coloured population is as valid as a 'racial' or 'migrant' explanation. This is a salutary reminder that in these same urban districts there is a much greater proportion of native-born whites. This economic interpretation can be extended by examining the position of migrants in the economic structure of the city. We saw above that in the 1950s and 1960s West Indian immigrants responded to the employment market and its requirements for low skill labour. At worst unskilled West Indian immigrants joined a pool of labour, most of which was from the host society and traditionally occupied the central locations in the metropolis, in which the hope and opportunities for work were far greater. Much of this pool came to form an underclass – of whatever colour or origin – which has sat at the base of the city's socio-economic structure. Its members have been caught in a circle of poverty, deprivation and semi- or complete unemployment from which it has proved extremely difficult to escape.

Preserving Identities

Peach (1968) made the point forcibly that identity can at times be preserved through economic replacement. Up to the Second World War, for example, Welsh migrants had a quite disproportionate role in the milk distribution trade in London. They virtually dominated the sector by the 1880s, as Booth's (1903) findings testify, making a success of it because of their willingness to work for less returns than the host population, as well as being prepared to work longer hours. The result was that prior to the Second World War, after which the emergence of larger milk companies robbed this sector of its individualism, over half the small dairy shops in London were in Welsh hands. The nearest equivalent to this phenomenon today is the takeover of the corner shop, often combined with post office facilities, by families from the Indian sub-continent. Unpublished research undertaken at the London School of Economics shows clearly how the more modest financial demands of such Asian immigrants enables them to survive economically, usually with the assistance of family labour and using accommodation over the shop for living. This is a very familiar pattern in Southall. One thing which adds to the success of such groups is that they are attuned to the needs of the local ethnic community. They offer ethnic foods and additionally act as meeting places and community centres. Such enterprises also find a niche outside the ethnic arena, with the corner mini-market or tobacconist run by an Asian family being a familiar component of otherwise white suburbs.

Outside the coloured communities reference has been made so far only to the Irish. There are other ethnic groups in London which merit consideration, including the Jewish population. Today, the designation 'Jewish' encompasses few who are not native born, and consequently the nature of the data is different from that considered above. Jews merit no separate category in the census tables, for visibility is much less than it was once, and assimilation has been rapid. In fact this group must be dealt with as self-identified, for much of the data derives from membership of Jewish congregations (Lipman, 1954, 1964). This being the criterion, we must disregard all those descended from immigrant families who have become secularized or who have in some way abandoned this religious community. Historical studies have shown that there was considerable movement from the original hearth directly east of the City of London even from the first wave of migrants in the 1840s. As their wealth increased lower middle-class Jews moved north from Islington and Barnsbury by the 1860s, with the wealthier taking the further step of moving into Bayswater. These areas then become thresholds for further movement. In the north, Highbury, Finsbury and Stoke Newington became popular, while by the 1890s the more prosperous stream expanded north-westward from Bayswater through St John's Wood to Hampstead, with an extension to Golders Green by the inter-war period. In the following decades, Jews became a conspicuous element in Hendon and Edgware (Jones and Eyles, 1977). Meanwhile the original hearth was greatly reinforced by immigrants in the late nineteenth century and this has remained a sector of concentration. But the most prosperous Jewish population is now centred in Hendon and forms an unmistakable element of the social geography of north-west London (Waterman and Kosmin, 1987). In all, 17% of London's Jews are in the LB Barnet, with 60% of these in just six of the 20 wards in that borough. Yet even with this concentration, in no ward do Jews form a majority, and only in Edgware do they make up 40% of the population. However, there is a tendency for clustering to occur on a street by street basis, giving small but high concentrations without formal segregation from the rest of the population. Waterman and Kosmin (1987) suggest that the dispersal of these small nuclei is a reflection of the local homogeneity of house sizes and types. In addition, the cohesion of the loosely assembled Jewish neighbourhoods reflects the need to be within reach of a rabbi, of kosher food, and to be within walking distance of a synagogue.

Relations with the Host Population

Two major questions arise concerning the role of ethnic and racial populations in London. One concerns the relationship between the host and the migrant community and the attitudes and behaviour that result from this. The other is the way the metropolitan community as a whole is seen in the future; viz. whether it will be integrated or pluralist. For two decades or so before the 1970s the growing number of migrants was perceived as a social problem which could be solved by a rational, liberal approach, guided partly by the ethnic experience of the United States. Any tendency towards the concentration of coloured people brought spectres of the ghetto, fears of social division and images of social confrontation. The ideal was held to be assimilation and consequently the dispersion of immigrant populations (Deakin and Cohen, 1970), on condition that the latter was voluntary. However, in this regard, experiments in the Netherlands of the dispersal of East Indians proved a failure. This is not surprising perhaps, for embodied within the ideals of both dispersion and assimilation is an assumption of tolerance in the host society, and that any contrary attitudes can be changed by education or at least can be controlled by legislation.

Not surprisingly, given the simplicity of these assumptions, notions of assimilation have remained as ideals with little indication that they will be realized. Indeed, in the absence of strong policy initiatives from central government, or even of agreement on strategy, there has been a mixed response from local authorities. This is critical, for in the realm of residential segregation, London's boroughs have controlled a vital element of assimilation; namely, access to housing. Some councils have discouraged the congregation of coloured migrants, even though they have acknowledged that it is probably unwise to separate widely people of similar origins. Other councils have used selection processes which have maintained the white character of certain of their estates (Commission for Racial Equality, 1984; Phillips, 1987). The adoption of this latter policy stance in no sense carries an automatic tag of an underlying rationale of racism. Many people recognize the benefits of congregation for immigrants, particularly when immigrants are still arriving in large numbers; the advantages of 'urban villages' are that they act as cultural cushions which enable newcomers to adapt themselves to a new life in an environment in which some things at least are familiar.

However, vague notions of eventual geographical integration failed to come to grips with the reality of the increasing antipathy of the host population. Indeed, that set of attitudes that constitutes racism only became fully appreciated when immigration increased considerably. Hence, while geographical concentrations of immigrants can be explained in economic terms, the rise of racism put the matter of eventual redistribution in a new perspective (Jackson, 1987; Smith, 1987). Certainly racism was exacerbated by the size of the coloured population, but more than this has been involved, for it is evident that racist attitudes are deep-seated, operating not only at a personal level but also at an institutional one. As Doherty (1973) made clear, racial distinctiveness is merely reinforced by socio-economic conditions; even in the socio-economic classes in which immigrants found themselves, they have been looked upon as competitors for scarce resources and, consequently, have been seen as a threat to the opportunities which are open to the white population.

Since the idea was introduced by Boal (1970) some two decades ago, a conflict interpretation has dominated interpretations of ethnic and racial issues in Britain. From this perspective, it is not surprising that the tense, often threatening, situation in which immigrants find themselves has had the effect of strengthening their communal identity. In London, as in many other British cities, this tension has led to open conflict on numerous occasions, which, in turn, has highlighted the existence of racial prejudice. Thus, a 1981 Home Office survey found that

Asians were 50 times more likely to be victims of racial attacks than whites, and West Indians 36 times more likely (HMSO, 1987). As Lord Scarman's report on the Brixton riot of 1981 made clear, the insecurity of West Indian youths provided an unstable context for social calm when intensified by police saturation searches (Clare, 1984), in a longer-term context in which area residents were convinced of their being targeted for the institutionalized racism of police harassment (HMSO, 1981; Greaves, 1984).

Overall, there is every indication that racial prejudice is acute and that it is unlikely to diminish with time. Indeed, at times confrontational attitudes are reinforced by the behaviour of coloured groups, partly out of a concern for their own defence and sometimes as an assertion of their distinctive identity. Many people with relatively recent immigration roots are not attracted by assimilation as an eventual means of circumventing such problems. Of course, the continued visibility of coloured groups to the host population lessens chances for integration, and adds to the attractiveness of laying stress on the positive attributes of immigrant identity, favouring a movement toward ethnic pluralism and cultural heterogeneity rather than assimilation. Unfortunately, for the West Indian migrant both of these prospects do not hold out much hope for rapid advancement; on the one hand on account of their obvious visibility, on the other hand due to the very obvious sharing of many aspects of immigrant culture with that of the host society. For West Indians, language has not been a barrier and neither has religion. Not surprisingly, therefore, one reaction to the perceived low standing of West Indian youth, and to their lack of a comfortable social niche within British society, has been an emphasis on less obvious cultural differences in order to create a new identity. Pursuing this path might not simply further distance this group from the host society, but could also separate them somewhat from the older generation within their own group. The clearest example of such a tendency is found in Rastafarianism, which is the rebirth of a creed which had its roots in the West Indies in the 1930s, but which refers back to the Coptic religion of Africa. Visibility here is reinforced by hairstyle, headgear and language; English gives way to a patois based on West Indian dialects, with some cockney elements, to produce a form of expression that provides a refuge from outside authority which is often called on in conflicts with whites (Sutcliffe, 1982; Pilkington, 1984). Reggae music provides one further element in the attempt to establish a distinctive cultural identity.

For other coloured populations there is no need for such a self-conscious cultural identification. As in the case of Asians, here cultural elements are commonly at variance with the host population from the outset. Indigenous languages are dominant, and religions are more inclined to be those of Islam, the Hindu or the Sikh. Lifestyles confirm the separate identity of these population groups, with cultural identities being kept fairly intact over long periods. Indeed, even when there is cultural seepage, as members of these coloured groups become more absorbed with the host population and as more of their members have moved into white neighbourhoods, the original geographical concentrations of such populations have often been reinforced by the addition of new immigrants. This is particularly true of the most recent, and the most disadvantaged of London's immigrants, the Bangladeshis, who have created a very distinctive population concentration around Spitalfields, to the east of the City of London (Runnymede Trust, 1987). This group faces a most difficult adjustment, for most come from the rural Sylhet region of Bengal, and so have a rural–urban adjustment to make in addition to a cross-national one. With a reported three-quarters of this Bengali population who are 15 years old having no fluency in English (HMSO, 1987), language is also a problem. These people look likely to be open to the prospect of severe discrimination, made all the more poignant by their place of residence, which occupies that part of London that has a long history of racial prejudice and overt, and often violent, reaction to immigrant minorities (Husbands, 1982).

There is no doubt that a sense of continuity of culture is strong amongst Asian populations. This has been reinforced by recognition in the symbols of their culture. The teaching of some Asian languages in inner London schools provides one example of this, with additional educational funding providing some reassurance of national recognition of their culture. More outwardly visible symbols, such as the temple, the mosque, and specialized food shops as conspicuous elements of the townscape, also help transform the scene in more closely knit neighbourhoods. Even more personal symbols, such as the wearing of turbans and head-scarves for females, have seemingly found a wider acceptance; albeit their full recognition across all walks of life might be a longer-term struggle. Indeed, it cannot be assumed automatically that the gains of the past are permanent. The abolition of the Inner London Education Authority, the local educational authority which has done much to provide a favourable environment for minority group education, is a case in point, for the language and other gains of the past must now be assumed to be under some threat as pressures for spending cuts are pressed on its replacements.

Conclusion

A glance at the future is bound to reflect everything that has been said on the relationship between host and migrant. Hopes for assimilation disappeared some years ago, and the socio-cultural strength of some of the coloured populations demands that we think in terms of continued diversity. There are degrees of acceptance over cultural differences on both sides, and the meanings of some of the terms used in this chapter may be helpful in clarifying the issues. *Assimilation* refers to the complete absorption of a migrant group and its apparent disappearance over time; the Huguenots offer a good historical example. *Integration* is a less complete process, though it implies full participation in and acceptance by the host society; the Jews may well fit into this category. *Pluralism* in many ways is the opposite of integration, for it implies a largely separate identity which is based on the acceptance of recognized differences. In such cases, high geographical concentrations of home places indicate a predominance of interaction within the migrant community; as is evinced amongst the Bangladeshis in East London. Then there is *separatism*, which is based on the total rejection of host values by the migrant group, with an attendant emergence of antipathy and confrontation. Possibly Rastafarian groups reveal some characteristics of this, although under different conditions they could be part of a pluralist society. Today, many would hold that pluralism is a desirable goal, but only on the basis on equality and tolerance. If these criteria find broad acceptance, then this might well lead to a greatly enriched metropolitan culture.

11

The Borough Effect in London's Geography
Michael Hebbert

Introduction

Though geographers often use the word 'London' in an elastic sense, the outstanding characteristic that distinguishes our capital city from New York and Tokyo is its existence as a hard geographical fact. Built-up London is a slightly irregular Rohrsach blot with extremities at Enfield, Hornchurch, Purley, and Uxbridge. It is surrounded and shaped in all directions by an impermeable belt of agricultural and recreational land. Its physical edge, unchanged along much of its length for 50 years, corresponds – give or take a few miles and some anomalies – with the boundaries of Essex, Hertfordshire and Buckinghamshire to the north and Berkshire, Surrey and Kent to the south. When memories of its suburbanization and the subsequent adjustment to shire county boundaries were still recent, the area within used to be called 'Greater' London. It is today's London.

The interesting aspect of the geography of modern London is not its indeterminacy but on the contrary, its clarity of definition – or rather, the paradox that so clearly defined an urban unit should lack administrative and political expression. The county of London is a void in the institutional map of Britain. It has no institutions of its own save for the little-known, feudal figure of the Lord Lieutenant, accompanying or deputizing for the Queen on ceremonial occasions within the capital. Its seat of government, County Hall, was sold-off to a short-lived consortium of financiers who tried to redevelop it as a luxury hotel. The physical unity of London is parcelled up into 33 separate areas for purposes of local government. Other chapters in this book have looked at London as a whole. Here, we focus instead on the separate pieces of the London jigsaw, and consider first the principles on which they were divided up, then the implications of the division for the geography of our metropolis as a whole.

London government is essentially *borough* government. The 31 London boroughs with the cities of London and Westminster are directly responsible for almost the full range of local government functions. A further set of services, of wider geographical scope, such as the fire brigade, is provided indirectly through agencies under the joint control of several or all boroughs. Lastly, a small number of politically sensitive and high-spending services lie under the direct control of central government ministries, some administered by civil servants, others through arm's-length agencies run by political appointees (see Hebbert and Travers, 1988). Because central government is, by definition, concerned with the nation as a whole, it tends to subsume London issues into those of the wider South East of England, as well as compartmentalizing them into the concerns of the various ministries. So although central intervention is all-pervasive in London, it is also rather diffuse, reinforcing the status of the boroughs as the only governments directly responsible to and for Londoners.

The Making of the Borough Map

The present pattern of borough government dates from the 1963 London Government Act, implemented just a quarter of a century ago (Figure 11.1). This reform achieved two things. First, it pushed out the administrative boundary of London so it more or less matched the suburban limits of the built-up area. The reorganization to 'bring London government into harmony with the physical features of the metropolis' (MHLG, 1961, 4) was a remarkably bold geographical gesture, resulting in perhaps the largest suburban incorporation ever carried out by a major city. Some political compromises were made in the reorganization process – Epsom and Ewell, for example, were taken out of London and restored to Surrey by a vote in the House of Lords – but they left intact the fundamental notion of London as a single geographical entity with its own system of government, originally a double-decker system, today a simple division into 33 parts.

The second effect of the reorganization was to reduce the number of units within London. The reform came at a time when expert opinion in the field of public administration in Britain was firmly committed to large units. The 1957–60 Royal Commission on Local Government in Greater London recommended a structure of 52 new boroughs with a population range of 100 000 to 250 000 (Herbert Commission, 1960). The government of the day decided that the primary building-blocks needed to be still bigger, and took 200 000 as the minimum population wherever possible. It believed that big boroughs would

Figure 11.1 Local government in London

attract a better calibre of councillors into local government. Also, '. . . larger units would mean more work for each authority . . . and so make specialisation in staff and institutions more efficient and economical. In addition larger units would be stronger in resources and so better able to secure the major redevelopment which many boroughs now need' (MHLG, 1961, 3). This decision meant the dissolution or merger of every existing authority in London except Harrow Municipal Borough and the City Corporation. Smaller councils such as Finsbury (population 35 400) put up a spirited and, in retrospect, highly persuasive case for their survival. But the times were against them.

Setting aside the inner/outer contrasts which were as strong in 1965 as they are today, the newly formed boroughs shared an important characteristic. Except Harrow, they were all compounds of previously neighbouring (and therefore rival) councils, and some of the mergers were bitterly opposed. The government delegated to the town clerks of Plymouth, Cheltenham, Oxford and South Shields the final intensely controversial decisions about the borough map. The case which gave them the greatest difficulty was the amalgamation of Wembley and Willesden, physically ill-matched partners in the new borough of Brent. Other notably unpopular mergers were East Ham with West Ham (LB Newham), Bromley with Beckenham, Penge, Orpington and half of Chislehurst and Sidcup (LB Bromley), Romford with Hornchurch (LB Havering), Hornsey, Tottenham and Wood Green (LB Haringey), and on London's western flank, the largest amalgamation of them all, combining the Urban Districts of Yiewsley and Drayton, Hayes and Harlington, Uxbridge and Ruislip-Northwood into the London Borough of Hillingdon. Gerald Rhodes (1970, 153) described these as marriages of administrative convenience: '. . . they may form workable patterns for running education, welfare or public library services but it would be hard to find any consistent reasons for the particular amalgamations which were finally accepted'.

Drawing the lines of the borough map was only half the battle. The units had to be named. The Minister, Sir Keith Joseph, decided to override the general desire of the old councils to keep their names and enforced a new and simple borough nomenclature, allowing only one hybrid name, Kensington and Chelsea, and that reluctantly. Dagenham (population 114 000) lost its identity to a lesser neighbour, Barking (75 000), and Fulham (113 000) lost its identity to Hammersmith (109 000). In the latter case the councillors considered 20 other suggestions before falling in with Keith Joseph's recommendation to call the new borough Hammersmith (Rhodes, 1970, 202). Neither Dagenham nor Fulham members forgot their old loyalties, and their boroughs were eventually renamed, with much repainting of signs and re-printing of letterheads, in 1981 and 1978 respectively. To avoid such difficulties in cases of amalgamation of evenly matched units, the new authorities were named neutrally after more or less obscure features (Hillingdon, Brent, Havering, Redbridge, Merton) or given synthetic titles (Tower Hamlets, Newham, Waltham Forest). But where there was already a dominant authority within the new borough boundary, the reorganization meant in effect the takeover of lesser neighbours. So, Finsbury lost out to Islington, Battersea to Wandsworth, Erith to Bexley, Penge to Bromley and Barnes to Richmond.

The pre-reorganization pattern of London government, which troubled reformers because of its irregularity and fragmentation, had fitted the topography of the metropolis like a glove. It matched the pattern of postal addresses, bus-destinations, street signs, local newspapers. Local councils corresponded to localities, and almost every significant sub-centre in the London metropolis had a town hall and local government to its name. The new units were more artificial and less tangible. A quarter of a century after the reorganization, Londoners carry Hornsey and Hornchurch in their mental maps ·of the metropolis rather than Haringey and Havering. Real maps – particularly the Ordnance Survey sheets for London, on both 1:25000 and 1:250000 scales – still prefer to identify districts by names corresponding to pre-1963 local authorities, making almost no use of borough nomenclature. Yet the modern division of London into 33 parts cannot be an unimportant geographical factor. The fact that London is the only world city divided up in this polycentric manner gives added interest to the task of identifying its effect on the geography of the metropolis. That is the purpose of the present chapter.

In Search of the Borough Effect

The wrong place to begin our enquiry is with choropleth maps of socio-economic indicators for the London boroughs, striking though some of the contrasts are. London's 33 areas tend to scatter to the extreme ends of national scales of comparison. To take one example at random, Newham and Havering are near-neighbours, but at the 1981 census one had England's highest percentage of children living in poor housing, the other the fifth lowest. Borough-based statistics, however, conceal much local variation, for the units are large and London's social geography is diverse, with some mixing of housing tenures in almost all parts of the metropolis. Maps of borough indicators raise an expectation of precipitous contrasts along boundary lines which are rarely fulfilled on the ground. Indeed, until the fashion for erecting boundary signs in the mid-1980s (an unexpected side-effect of GLC abolition) it was often difficult for a non-resident to know where one borough joined the next.

Variations in the *overall* socio-economic and political composition of the boroughs do, of course, imply significant inter-borough differences in taxes and services. Where the boroughs of Croydon, Bromley, Lewisham, Southwark and Lambeth converge by Crystal Palace, a household may find itself with a tax bill double that of its identical neighbour. The London boroughs include some of the highest-taxing local authorities in the country and some of the lowest. The range was exaggerated in the early 1980s by expansionist 'new left' politicians who financed programmes out of the rates in the knowledge that the welfare system would shield those on council estates from additional fiscal burden. It narrowed later in the decade after the government passed legislation to restrain levels of local property taxation (rate-capping), but is expected to widen again after the introduction of the local poll-tax in 1990.

The borough effect will be amplified in 1990 by the abolition of the Inner London Education Authority, as each inner borough will become an education authority in its own right. School catchments are now being realigned to fit borough boundaries, so that cross-boundary traffic in pupils is regulated and charged to the appropriate authorities. As educational standards diverge, some boroughs being wealthier or administratively more competent than others, access to schooling could enhance or diminish the relative attraction of inner boroughs as it already does those of outer London, which have been educational authorities from their inception. However, the locational significance of educational differences should not be exaggerated. Only a minority of London households have school-age children, and 10% of these are educated in independent schools. What is more, the importance of local education authorities seems likely to decline under the 1988 Education Act. If schools were to opt out of the local authority system into direct grant status they would be able to operate their own selection procedures, making borough boundaries once again irrelevant.

What we do not see in London is the vicious political geography of the great American cities in which suburban governments, by geographical selectivity of population and real estate development, can provide a progressively higher level of service from a light tax burden, while inner districts move in the opposite spiral through a combination of rising demands on public service and a diminishing tax base. There is no evidence of a borough effect in London's geography in this sense of a fiscal polarization, working in a cumulative way to accentuate relative disadvantage. Thus, in 1987–88, households in Tower Hamlets contributed an average £501 for a council expenditure of £2100 per household, while in Kensington and Chelsea the average rate contribution was £524 for an expenditure of only £840 per household (CIPFA, 1989). Despite the switch to the poll tax, the local government finance system in London remains broadly redistributive, with central government contributing an average 70% of borough revenues in the form of grant and business rates. Of course, the centre's grant formula may favour certain categories of borough and penalize others. But these differences tend to vary from year to year and it is hard to prove that they have had a cumulative geographical effect.

Borough location is consequently not in itself a major factor in the social geography of London. The postal districts are more significant indicators of status. They follow a strange logic of their own, crossing borough and county boundaries and the 071– and 081– telephone areas, and because they define smaller, more socially homogenous areas, it is they and not borough names which carry the connotations of status and value equivalent to the Parisian *arrondissement*. In the property market, SW19 is perceived as a better location than SW20 though both lie within LB Sutton. Telephone codes are another more precise indicator of status which can exert a marginal effect on property demand: '736' is a 'good' code; it means a line connected to the Fulham exchange. '737', Brixton, is not so good. Local authority area counts for little in the London property market. *Where to Live in London,* a standard guide to property buying, is organized by postal districts, and makes hardly any reference to boroughs (Vercoe, 1988). Its rival *London Property Guide* gives them greater prominence – including lively thumbnail sketches of each – but its detailed purchasing guide is based on a division of London into 98 areas, 40 of which correspond to pre-1965 local government units and only one, Newham, to a modern London borough (Segrave, 1989).

The Importance of Planning

The lasting geographical imprint of the boroughs is not to be found in their taxation policies or their public services but in London's topography – the physical layout of roads, buildings and places. Local government shapes topography – and, on the larger scale, urban structure – by its building and public works activities and through the control of private development under the Town and Country Planning Acts. In modern London it is the task of the boroughs to draw up statutory plans for their areas and control new development. Since their inception, every borough has had a planning committee and a team of professional officers to analyze and understand its land use and shape its evolution, both by promoting change and averting (or diverting) it.

To appreciate the impact of borough decision-making on the structure of London we need to recall that these 33 councils have not had a free hand as planning authorities. In fact, the prime motive behind the reorganization of London government in 1963 was to create a single planning authority for the metropolis as a whole (Sharpe, 1965, 8–9). The 1963 London Government Act, for reasons explained in Rhodes (1970, 182–83), did not fulfil this purpose. The 33 boroughs instead acted as planning authorities in their own right, with the Greater London Council in an ill-defined supervisory role, providing strategic support and policy but with no power to approve or refuse borough plans. Inherent difficulties of a system of overlapping powers were compounded by a ratio of 33:1 between the boroughs and the GLC which precluded effective face-to-face communication between planners at the different levels. It did not help that many GLC planners and members came from a background in the LCC and tried to deal with the boroughs as minor, subordinate authorities, which legally they were not.

Much has been written about the battles between the two tiers and about the GLC's failure to achieve strategic hegemony, particularly over roads and public housing. Its road-building ambitions were crystallized in the initial proposals of the Greater London Development Plan for a comprehensive London motorway network of three ringways and 13 radial roads (GLC, 1969). A classic top-downward road engineer's scheme, it tackled the technical challenge of highway design but left all the problems and the (financial and political) costs of environmental alleviation to the boroughs, who had not been consulted in its preparation. They responded by joining amenity and community groups in a broad front of opposition to force the abandonment of the strategic road plan in 1973 (Hart, 1976, 158–75). So London had the good fortune to become the only major British city without a primary road network. Its absence is an aspect of the borough effect.

Another element of the GLC's original planning strategy which the boroughs blunted was its 'metropolitan structure' of broadly alternating land use – 'settlement areas', 'work areas' and 'metropolitan open land' – studded with six major subcentres at Ealing, Wood Green and Ilford to the north, and Kingston, Lewisham and Croydon to the south. The concept of a selective deconcentration within the original Greater London Development Plan had immediate parallels with metropolitan strategies for Paris and Tokyo, but with two crucial differences. Sub-centres such as La Defense and St Denis (Paris) and Shinjuku and Shiboya (Tokyo) were developed as major transport nodes. Their plans involved a degree of integration between land use and transport planning never achieved in County Hall. And whereas the metropolitan authorities in Paris and Tokyo had the political weight to promote their designated growth-points, the GLC was forced onto the defensive by rival claims from other borough centres such as Wandsworth and Bromley, and criticized by the inspector at the public inquiry into the GLDP for being over-intrusive in its attempt to impose a 'metropolitan structure' (HMSO, 1973). So the original, top-downward, conception of a strategy for London gave way to a pattern better fitted to borough interests, based not on selective promotion of six major centres but on a polycentric model of 28 strategic centres, (almost) one for each borough.

Another aspect of the GLC's strategic planning effort tempered by borough influence was housing policy. Strictly speaking the GLC was never intended to become a housing authority. However it inherited an enormous stock from the LCC and perhaps more importantly, a large team of architects and surveyors who gave institutional momentum to the housing programme. There was encouragement too in the Council's formative years from the 1964–70 Labour government which saw the GLC as the flagship of its national housing programme. The Council's moral justification was provided by the report of the Milner Holland Committee on *Housing in Greater London* (HMSO, 1965), in which the spacious densities of 20 persons per acre in suburban boroughs were contrasted with the 200 in inner London. Unfortunately, the drive to equalize housing opportunities for Londoners coincided with the worst period of postwar public housing design. Massively intrusive developments in grim concrete, like the Kidbrooke Park Estate to the south of Blackheath, made a poor advertisement for strategic planning. Suburban politicians used every legal, political and administrative means they could to frustrate such schemes, forcing the GLC housing machine to seek sites in Labour-controlled boroughs of inner London. They were obliged, sometimes against their inclination and better instincts, to accommodate the greater part of the GLC's output of 82 000 housing units in 1965–85. The GLC's inability to 'open up the suburbs' has become a celebrated case-study of failure in metropolitan strategy (Young, 1980; Young and Garside, 1982). It reinforced the dominance, already stronger in London than other British cities, of council housing in the inner areas, and it accelerated inner borough economic decline through the extinction of manufacturing activity within nineteenth century streets that were bulldozed for monolithic residential development (Buck, Gordon and Young, 1986, 12 and 56).

The bias of GLC housing policy was symptomatic of a more general weakness. Frustrated in its attempts to do what it had been set up for – laying down strategic policy for land use and transport across metropolitan London – the council redirected its energies into direct interventions which it could itself plan, finance and execute. The GLC became, as it were, a thirty-fourth borough, devoting its enormous financial resources to a rather miscellaneous group of projects that fell under its direct control (Self, 1972). As well as housing estates in inner London these included the opening up of two, still unfinished, major new parks in Tower Hamlets and Southwark, and the building of distant New and Expanded Towns, for which the GLC continued energetically emptying the capital's manufacturing base throughout the 1960s and most of the 1970s (Buck, Gordon and Young, 1986). In deciding where to focus its necessarily selective planning efforts the GLC had to take notice of borough opinion. Its attempts to initiate 'action areas' of large-scale planned redevelopment at Victoria, King's Cross and Piccadilly foundered on the hostility of borough councils for these areas. It was no more successful in the redevelopment of the docklands, where consultation procedures with the five riparian boroughs absorbed the best part of a decade before the government stepped in and placed the area under a centrally appointed *ad hoc* corporation. It succeeded best where it could work with the boroughs, as in its local road schemes around Ilford, Kingston and other suburban shopping centres, and the 17 areas in which it concentrated resources for housing rehabilitation and environmental improvement.

If the GLC failed to leave any coherent imprint on the geography of London that was because its action space was circumscribed from above by central government and from below by the boroughs (Self, 1972). Here was the strange paradox of London government in the years 1965–86. The Greater London Council had the prestige and the publicity. Its politicians saw themselves as playing in the first division and its team of two hundred planning and transportation officers had salaries up to 30% higher than their borough counterparts. Yet its actual powers proved, in Young's (1980, 24) words, 'ambiguous, temporary and unworkable'. Development control – the granting or refusal of planning permission for building works or changes in the use of a property – lay in the hands of the boroughs in all but a small number of special cases (0.95% in 1983), though these cases were the important ones and the GLC's power to 'direct' the decision upon them was much resented. In policy-making, it was the boroughs and not the GLC that took the lead in metropolitan planning. Many of the policies set out in the approved GLDP of 1976 were generalized and permissive. The Panel of Inquiry famously observed of them that '. . . the GLDP Written Statement is full of statements of aims which do not mean anything because they can mean anything to anyone. It is not perhaps being too cynical to believe, indeed, that such aims were inserted because they could mean anything to anyone' (HMSO, 1973, 27). That flexibility was chiefly advantageous to the boroughs, whose planners devised and implemented specific residential density policies, plot ratio formulae, industrial and commercial floorspace targets, car-parking standards, and retail zones. Over the past quarter century these have shaped the concrete morphology of London. Yet some provisions of the GLDP have been widely employed by boroughs in support of local policies: for example, the plan's protection for 'metropolitan open land', and its policy of concentrating shopping, office and factory development in 28 'strategic centres', 40 'preferred locations for offices' and 43 'preferred locations for industry'. The GLDP was effective insofar as it could be so used. Likewise, GLDP proposals that did not run with the grain of borough priorities rarely progressed; that was why fewer than half of the 54 'action areas' proposed in the GLDP resulted in action (Field, 1983).

The weight of borough policy-making became fully evident when the GLC in its final years updated the strategic plan for London (GLC, 1984). Probably the most important GLDP modification proposed in 1984 was the distinction between a 'central activities zone' – for functions associated with London's regional, national and international roles – and a surrounding belt of 'community areas' where the paramount objective was to safeguard existing residential communities from pressure for commercial development. This philosophy of CBD containment was drawn directly from the local plans of borough councils, both Labour and Conservative, who had sought to protect inner residential districts such as Fitzrovia, Bloomsbury, Soho, Rotherhithe and Battersea from redevelopment pressures exerted by the City and West End (Healey, 1983, 108–11 and 191–92; Bruton and Nicholson, 1987, 147–48). Even before the GLC was abolished, London's planning was driven, as it is today, from below and not from above. The question is, what have the boroughs been trying to achieve?

The Planning Objectives of the Boroughs

The obvious place to look for planning policies is in plans. Barrie Morgan does it for his analysis of borough shopping policies in Chapter 7. The quality of these documents is very uneven. The boroughs were given no clear encouragement or guidance over plan preparation; a change in the regulations in 1972 left them with no formal obligation to prepare one at all, and in more static parts of the metropolis the planning system could run quite adequately on the basis of precedent, informal policies, and the detailed development plans drawn up by the London County Council and eight adjacent authorities in 1954–58 (Field, 1982, 1983, 1984; Healey, 1983). The building boom caused by the upturn of the London economy in the 1980s made it more important for councils to be able to base their planning decisions on a formally approved and up to date plan.

Under the provisions of the Local Government Act 1986, each must prepare a 'unitary development plan' for its area. By 1992 the patchwork of these plans should cover the metropolis.

It might seem impossible to generalize about the planning efforts of the London boroughs. Like most local government business in the capital it has been affected by political posturing. Inner London boroughs under 'new left' control from the late 1970s set a trend of using town plans as manifestoes for declaratory redistributive and anti-discriminatory policies. The radical 'new right' council which gained control of Wandsworth in 1978 likewise introduced political posturing into its 1980 Borough Plan, which, after a tumultuous local public inquiry, prompted a plea by the Inspector for town planning to be kept neutral, so '. . . it may serve its true purpose without becoming a party political football to be kicked between opposing political goals' (Healey, 1983, 157). Yet despite the great variety in the policy styles and values of councils, and in their urban environments, I shall argue that there has been a latent common pattern in the planning policies pursued. Over the years this pattern has exerted a real geographical influence on London's geography – the *borough effect* of our title.

Defining the Borough Effect

The borough effect is a straightforward expression of political geography. The units created by the reconstruction of local government in 1965 had a new territoriality, distinct from their predecessors and competitive with their neighbours. Each arbitrary tract of built-up London defined in the reorganization process became for borough leaders and their officers a 'field of vision expectation and action' (Young and Kramer, 1978a, 238). Original marriages of administrative convenience became, for those at the centre, real entities worth campaigning for, with boundaries that showed on council wall-maps like an island shoreline. Political geography identifies a general tendency – at all scales of government – for the spatial development of an area to follow and reinforce the centre-periphery perspective of those who govern it (Gottmann, 1980). In the case of London we can hypothesize that a maturing territorial identity on the part of boroughs will have tended to find physical expression in the growth of their centres and the downgrading – or peripheralization – of 'failed' centres which had lost the status of autonomous local government areas. Twenty-five years is long enough for this physical expression of centrality, the borough effect, to leave its mark on the topography and geography of London.

Figure 11.2 shows the hypothesis diagrammatically. Three former local government units are merged under the 1963 Act. The new borough's civic centre is located in the larger of the three, which becomes – *de facto* – the heart of the new borough. Its shopping and services are designated as 'strategic' and its expansion encouraged through the council's own projects, in joint ventures with private developers, and by the exercise of development control. The other two centres become 'secondary' and their shopping and other services stagnate or decline. One of the former town halls is used for council office accommodation, the other demolished. So the initial administrative unification takes effect, over the passage of years, on the geography of the area, as three areas with their own centres are rearticulated into one with a single core. If the division into 33 units has had a structural influence on the geography of London, it will show up as a polarizing factor, concentrating activity in some centres, depressing it elsewhere. Has this occurred?

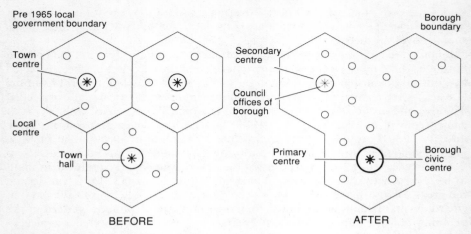

Figure 11.2   The borough effect

The Borough Effect – Concentration

The first clue to the borough effect is the location of council headquarters within the new local government areas. Reorganization in 1965 concentrated local government administration in a small number of large bodies. Not only did each council cover a substantially larger population but the trend of local government in the 1970s was relentlessly expansionist. More and bigger buildings were needed, and their location and design was (after the choice of name) the most important expression of a borough's corporate identity. And like the choice of name, the location of the seat of local government was a sensitive issue whose outcome depended partly on the strength of old loyalties. In the majority of boroughs, its outcome was a concentration of government in the heart of the largest pre-reorganization authority.

Municipal concentration could take various forms. Most simply, an annex could be added to the town hall in the main urban centre, as in Ealing, Enfield, Woolwich (LB Greenwich), Hammersmith, Wood Green (LB Haringey), Romford (LB Havering), Kingston, and Lewisham. An alternative, were such expansion was physically difficult or politically sensitive, was to create a modern civic centre on an extensive site nearby or on the edge of the dominant urban centre, as in Bexley, Bromley, Uxbridge (LB Hillingdon), Hounslow, Stratford (LB Newham) and Sutton. Croydon and Westminster achieved the same end by taking over conveniently located speculative office blocks. Whether by expansion, new building or leasing, two-thirds of the boroughs located the seat of local government in their dominant town centre – the 'strategic centre' defined in the Greater London Development Plan (GLC, 1976a, 80–1).

The dovetailing of the two was no accident. One of the foremost considerations in drawing borough maps 30 years ago was that each local government unit should be based on a substantial service centre. London's entire system of town centres had been meticulously researched by the geographer Ian Carruthers, who first helped provide the technical basis for the 1965 reorganization as research officer at the Ministry of Housing and Local Government (Herbert Commission, 1960, 295–305), then joined the GLC and wrote the section of the Greater London Development Plan defining the 28 'town centres of strategic importance' (GLC, 1969, 42–3), all but four of which (Peckham, Clapham Junction, Holloway and Kilburn) were also a seat of borough government, and all but one an established High Street.[1]

Though the original GLDP had said nothing about the prospect of commercial office development in 'strategic centres' (GLC, 1969, 18–21), the boroughs successfully lobbied for an office policy that would reinforce those for shopping and transport (see Glassberg, 1981, 169–70). So in the final version of the GLDP, all but a handful of London's 'preferred locations for offices' outside of the City of London and the West End were distributed equitably amongst the borough centres – Barking, Bexleyheath, Bromley, Croydon, Ealing, Enfield, Woolwich, Kingsland, Hammersmith, Wood Green, Harrow, Romford, Uxbridge, Hounslow, the Angel at Islington, Kingston, Brixton, Lewisham, Wimbledon, Stratford, Ilford, Richmond and Twickenham, Peckham, Sutton and Walthamstow (GLC, 1976a, 28–9).

The consolidation of a borough's core was often accompanied by urban renewal and road improvement projects. Strategic centre development was one of the points where the GLC and boroughs had a mutual interest, enabling large-scale schemes of urban remodelling to involve the four elements of office development, roads (and multi-storey car-parks), shopping centres and a civic centre, based on the already successful precedent of Croydon (GLC, 1969, 68–73). The risk of GLC involvement was that roads and car-parking dominated. The boroughs of Barking, Haringey, Newham and Greenwich had cause to regret the early redevelopment of Barking, Wood Green, Stratford and Woolwich with destructive, inhospitable schemes that have lost competitive position to more agreeable places and are now classified as problem centres (Table 7.1). Ealing and Hammersmith were fortunate to have similar redevelopments of their Broadways blocked for many years by community opposition until the wheel of architectural fashion turned back towards conservation and urban design, bringing Ealing its splendid Broadway Centre and Hammersmith its King's Mall and Bredero schemes. Other successful examples of strategic centre improvements include Bexleyheath, Bromley (still in progress), Ilford, Romford, Uxbridge, and Kingston. In all these cases large-scale retail development has been combined with High Street pedestrianization and landscaping, a mixture of old and new which enables them to compete successfully on quality with both the West End and large freestanding centres further out (SERPLAN, 1989a, 69–95).

Once the policy framework of strategic centres was in place, first in the Greater London Development Plan and then in the local plans of individual boroughs, it tightly channelled major retail and office investment. Planning permission was hard to obtain elsewhere, unless by wayward appeal decisions. Within the centres, developers could look to support and sometimes partnership with the local authority. Political will and market control combined to produce rapid topographical change. New concentrations of offices – some not well planned – appeared in Hammersmith, Ilford, Ealing and Hounslow in the 1970s, and in Bexleyheath, Hillingdon, Wandsworth, Richmond and Uxbridge, in the 1980s (LPAC, 1987a). The strategic centres have captured the lion's share of recent and forthcoming retail investment (LPAC, 1987; SERPLAN, 1989a, 1989c, 1989d). Following Redbridge's appointment of an Ilford Town Centre Manager with chief officer status, London councils have pioneered comprehensive schemes of management for the quality, security and competitiveness of their leading centres (SERPLAN, 1989c, 8). More recently still, under the stimulus of the *London Tourism Strategy* (London Tourism Board, 1987), they have begun to compete for the tourist dollar. By 1990, five boroughs had appointed 'tourism development officers', and over 50 potential sites had been identified for new hotel development, 12 in borough centres (Barking, Bexleyheath, Croydon, Ealing, Harrow, Ilford, Richmond, Romford, Lewisham, Uxbridge, Wandsworth and Wimbledon). Similar patterns might also be found in arts funding, street planting, and signposting. Boroughs have a strong implicit commitment to maintain their share of business against neighbouring authorities, and the main town centres are the flagships in this, generally amicable, trade rivalry.

The Borough Effect – Peripheralization

Boosting borough cores implies a downgrading of secondary centres. The downside of the borough effect is most obvious in the conduct of local government activity. In their evidence to the (unsympathetic) Herbert Commission, most of the doomed smaller local authorities argued that their town halls served as local enquiry points for citizens in their dealings with local government, including the rather wide range of services which these authorities were too small to supply. They were, as we now say, 'first stop shops', and in a very few cases have recently regained that role through borough decentralization initiatives (e.g. Acton, Poplar). But as borough administration consolidated, most of these former centres of civic life became peripheral. Only in Barking and Dagenham was a conscious attempt made to maintain parity between both of the borough's splendid pre-1965 town halls, alternating council meetings and distributing chief officers equally between the two sites. More common was the downgrading of former town halls to office accommodation for the client-based council departments – typically, social services or housing. Some have been turned into registry offices (Barnet, Chelsea, Hornchurch, Finsbury, Southwark), educational establishments (Bexley, Friern Barnet), crown courts (Beddington and Wallington, Surbiton), or premises for voluntary or arts organizations (Battersea, Finchley, Finsbury,

Table 11.1   Pre-1965 town halls since sold or demolished

Year	Town Hall (Present Borough)
1965–79	Barnes Town Hall (LB Richmond)
	Greenwich Town Hall (LB Greenwich)
	Paddington Town Hall (City of Westminster)
	Sutton and Cheam Municipal Offices (LB Sutton)
	Willesden Town Hall (LB Brent)
1980	Feltham Council Offices (LB Hounslow)
1982	Orpington Civic Offices (LB Bromley)
	Kensington Town Hall (RB Kensington and Chelsea)
1983	Ruislip/Northwood UDC Offices (LB Hillingdon)
1985	Heston and Isleworth Town Hall (LB Hounslow)
1986	Westminster Town Hall, Cavell House (City of Westminster)
1988	Beckenham Town Hall (LB Bromley)
	Drayton Hall (LB Hillingdon)
	Malden and Coombe Council Offices* (RB Kingston)
1989	Edmonton Town Hall (LB Enfield)
	Wanstead and Woodford Municipal Offices (LB Redbridge)
	Wimbledon Town Hall* (LB Merton)
1990–	East Barnet Town Hall* (LB Barnet)
	Crayford Municipal Offices (LB Bexley)
	Deptford Town Hall (LB Lewisham)
	Marylebone Council House (City of Westminster)
	Southall Town Hall (LB Ealing)

*demolished, façade retained

Mitcham, Richmond, Southgate, West Ham). A surprisingly large number have recently been sold off or demolished (Table 11.1).

For many districts, the loss of the town hall was echoed by a downgrading of shopping and other services. Ian Carruthers's work for the Herbert Commission (1960) and the *Atlas of Greater London Shopping Centres* (Thorpe and Kivell, 1974) together give a full picture of the position of all town centres in London prior to reorganization in 1965. We have no exactly comparable data for today which would allow a systematic comparison of the rise and fall of centres in the past quarter century. But two trends are perfectly clear. The first, analyzed in this book by Barrie Morgan, is the relative decline of all 14 centres in inner London. But cutting across the familiar inner/outer contrast is another trend, part of the borough effect, namely the relatively poor performance of district centres. The data, though patchy, suggest that secondary centres have higher levels of vacancies, a lower (and declining) share of multiple stores, less pedestrianization and parking, and less investment in the pipeline (LPAC, 1987; SERPLAN, 1989c, 1989d).[2] Secondary and tertiary centres have also been worst affected by the rise of out-of-centre shopping, although less has been built in London than in the rest of the country, thanks to the uniformly restrictive attitude of borough councils. 'Hypermarkets are being advertised as providing the whole range of High Street shopping under one roof. They are effectively adding whole new district level centres to the shopping hierarchy ... It is this level of centre that is already particularly vulnerable in London' (LPAC, 1987, 92). Although opinion surveys amongst retailers reveal a strong commitment to the traditional High Street, chain stores have been steadily trimming their smaller branches (Property Market Analysis, 1988).

There is evidence that the general pattern of peripheral decline is accentuated in the case of high streets that are physically on the edge of boroughs, or – as occurs in at least seven cases – straddle two or more boroughs (Table 11.2). Finsbury Park is an instance of the penalties of peripherality. It lies on the borders of Hackney, Islington and Haringey. It is highly accessible at the junction of four major roads, with a major transport interchange between two tube lines (Victoria and Piccadilly) and the Great Northern suburban rail services. Yet this centre has 10% of its shops vacant or derelict, no large foodstores and hardly any multiples (Rendel Planning, 1988). The effect of physical peripherality seems

Table 11.2 Town and district centres on borough peripheries

Centre	Boroughs (* = adjacent)
Balham	Wandsworth, Lambeth*
Chiswick	Ealing, Hammersmith and Fulham, Hounslow
Cricklewood	Barnet, Brent, Camden
Crystal Palace	Lambeth, Croydon, Southwark*, Lewisham*, Bromley*
Deptford	Lewisham, Greenwich*
East Acton	Ealing, Hammersmith and Fulham
Finsbury Park	Haringey, Islington, Hackney*
Herne Hill	Lambeth, Southwark
Kilburn	Brent, Camden, Westminster
Leytonstone High Rd	Newham, Waltham Forest
Norbury	Croydon, Lambeth
Southall	Ealing, Hillingdon*
Streatham	Lambeth, Wandsworth*
Tottenham Court Road	Camden, Westminster
Tooting Broadway	Wandsworth, Merton*
Waterloo	Lambeth, Southwark*

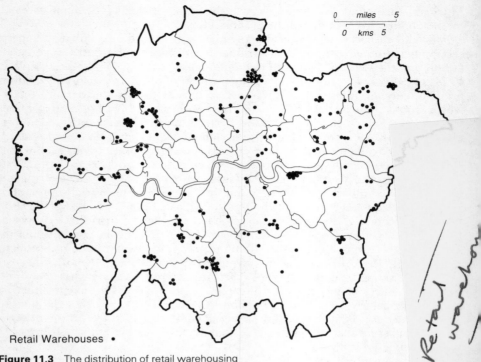

Retail Warehouses •

Figure 11.3   The distribution of retail warehousing

to be powerful. None of the edge-of-borough centres listed in Table 11.2 has performed well for its size (LPAC, 1987, 47–55; also Chapter 7). Such a pattern has a rationality: why incur the political costs of new retail development and the expense of environmental improvements when the benefits will flow over the border?

Further and positive evidence of a physical periphery effect is found in borough treatment of retail warehousing – the 'second wave' of the contemporary revolution in shopping. Like most new retail developments, the rise of operators such as TEXAS and MFI caught the planning system unawares, and it was some time before boroughs had policies in place to meet the demand for warehouse units with parking. The existing distribution of units (Figure 11.3) is the product of some initial *ad hoc* policy-making and appeal decisions, with a recent firmer policy to group these ungainly grey boxes in designated 'retail parks'. The clustering along borough boundaries is not coincidental. Partly attributable to historic river and road lines – which made natural local government boundaries and locations for industrial and warehousing estates – this clustering also reflects explicit planning efforts to accommodate retail warehouses where they are least likely to draw trade from the primary centre; in other words, the borough effect.

Much has been written about the contrast between inner and outer London. In this chapter we have been looking at centre-periphery contrasts at the borough scale, and at evidence of backwash from the boosting of borough centres. There was no historical inevitability about these trends. It can be argued that they counter the national trend for the strongest growth pressures to bear upon medium sized centres (LPAC, 1987, 104). Political concern at the relative decline in accessibility to shopping and services for the non-driving majority of

Londoners prompted the Greater London Council in its last years to replace its earlier policy of concentration on 28 strategic centres with an absurdly diffuse listing of 115 town centres (GLC, 1984, 173–177). The policy switch was justified on the grounds that in some boroughs there had been an 'excessive concentration of facilities in one centre, weakening others'. The plan went on to state:

> The uneven success of certain established centres in London introduces significant inequalities in the availability of facilities between different areas. The Council will give priority in its own programmes to those areas where deficiencies are most serious ... and will consult with the London borough councils over improving the environment in already well-developed centres rather than adding to facilities there. (GLC, 1984, 173)

Had the plan been adopted there would certainly have been tension between the GLC and the boroughs over the operation of this policy. But instead the GLC was abolished and, through the London Planning Advisory Committee (LPAC, 1988, 31–2), the boroughs have reaffirmed the view, not shared by government (DOE, 1989a, 21), that major new retail investment should have its 'principal focus' in 33 strategic centres. The borough effect seems set to continue.

Qualifying the Generalization

The observant reader who knows London will have noticed that we have so far skirted around the cases that do not fit the pattern. Spatial reordering around a single dominant centre has not occurred in all boroughs. Central functions may be split between two or more core areas. In LB Richmond upon Thames – the only borough to straddle the Thames – Twickenham has the council headquarters but Richmond is the major shopping and tourism centre and performs the role of the borough's eponymous central place. LB Newham also has a split core, with council headquarters in East Ham Town Hall at the further end of the borough from the main shopping centre in Stratford. LB Hackney has three equal centres at Dalston, Stoke Newington and Mare Street. Contrary to our hypothesis, the presence of the town hall in the third of these seems to have had no effect on its relative centrality, though this will change if current plans are implemented to develop it and the adjacent Hackney Empire, as a civic complex with new shopping. The least-focused boroughs are Barnet and Tower Hamlets. The former has the huge freestanding Brent Cross shopping centre, a scatter of prosperous but modest suburban High Streets, and an out-of-centre town hall at Hendon. Until the Isle of Dogs development is completed, Tower Hamlets has no High Streets to speak of and no new centres either, though there have been many projects over the past 25 years, including a grand scheme for a new civic core in the heart of the borough at Mile End. No doubt connections can be made between the lack of a central place in this borough, and its current experiment in radical political and administrative decentralization to neighbourhoods.

Another notable exception to the borough effect, as here defined, is the case of LB Brent. Kilburn, its largest centre, straddles the borough boundary, leaving Wembley as natural candidate for the borough's central place. Unfortunately it had lost its civic role when the former council moved in 1939 to an out-of-centre town hall built on Forty Lane, where LB Brent now has its headquarters. Wembley's role as a shopping centre has also been severely affected by the proximity of Brent Cross. It has fallen steeply in the ranking of London town centres by turnover; several multiples have reduced their size – C & A by 50%, Marks & Spencer by 35% – and many have ceased trading altogether (Dorothy Perkins, Burton, Hepworth, W H Smith). Wembley's plight mirrors the public perception of exceptional disintegration in political life and public services in LB Brent, and again prompts the question, not to be pursued here, whether there are connections between the absence of a topographical 'borough effect' and civic failure.

Conclusion

The geographical literature on modern London has a great deal to say about the late Greater London Council but rather little about the boroughs (see Dolphin, Grant and Lewis, 1981). This chapter has tried to correct the balance with a bold hypothesis, that boroughs rather than the GLC have exerted the main public influence on the pattern of development in London, and that borough policy has tended to favour cores over peripheries. The borough effect is more readily visible in suburban boroughs, with their few and far-spaced town centres, than in the denser urban environment of inner London, where centres merge one into another along linear shopping frontages. It has also had more purchase in a context of economic growth than decline. Nevertheless, we can argue as strongly for its existence in inner as in outer London. Central London would have had a different appearance today had it not been located on the peripheries of at least seven boroughs. Perhaps the most powerful manifestation of the borough effect can be seen directly across the river from the King's-LSE Joint School of Geography, in the new estate of low-density housing that occupies the prime development site between Waterloo Bridge and Blackfriars Bridge. Why is it there? The area was identified in the GLDP as a 'preferred location for offices' (GLC, 1976a) but was zoned the next year by LB Lambeth (1977) for housing and industry, with the rubric 'future office development to be resisted'. Three miles to the south, in the exact geographical centre of the borough, lie the Town Hall and Brixton town centre, where major office development is sought (LB Lambeth, 1984). Brixton is the core of Lambeth, Waterloo its northern periphery. The spectacle of a quasi-suburban housing scheme in the heart of London should startle any urban geographer. Whether we think it good or bad, it shows the borough effect is a force to be studied.

Notes

1   The exception was Brent Cross, which was '... planned to be a centre of the North American style for people who wish to use their cars for shopping' (GLC, 1969, 42). Carruthers' original analysis of service areas had revealed a 'gap' between the major centres at Wood Green and Ealing in the north-west of London. Brent Cross was a strategic planning initiative to fill the gap (GLC, 1969, 44), and remains, for the time being at least, the exception to prove the rule that London's retail modernization has been accommodated *within* and not outside its leading town centres.

2   Some 'peripheralized' centres, because of their lower rentals, have offered flourishing entry points for immigrant shopkeepers and so have both low shares of multiples and low vacancy rates. Tooting, Southall, Shepherd's Bush and East Ham are cases in point.

12

Rethinking London Government
Robert J. Bennett

Introduction

How is the government of a world city that is the size and importance of London to be organized? What weight is to be placed on the needs of the people who live and work in London compared to the requirements of the international market which shapes the demand that supports most of its jobs? How are the functions of London as the nation's capital – in terms of administration, politics, financial markets, media and culture – to be balanced with the needs of London's local communities? How is the weight of different local community opinions on development and other matters to be balanced between London's different communities in its highly integrated and complex labour market?

These are the questions that those responsible for the government of London have wrestled with over time. With the establishment in 1889 of the first integrated metropolitan government – the London County Council (LCC) – London government offered a paradigm for integration which was discussed the world over. This pattern was reformed in 1965 to establish the Greater London Council (GLC), which was abolished in 1986. Since abolition, despite the gloomy political predictions, no breakdown of services has resulted, few difficulties have emerged except in the fields of transport, and the concept of abolition of metrogovernment is on the agenda in other countries. Clearly a change of paradigmatic proportions has taken place.

In this chapter this change is analyzed over four dimensions: the evolution of the government of London as a world city, as national capital, for the metropolitan area and for local needs. A long term perspective is used which takes the present governmental structures as a starting point, but analyzes these, first, in the context of past arguments that have stimulated reforms, and second, in the context of current thrusts for reform that are seeking to address the governmental requirements perceived for the future.

One theme overarches all that is said in this chapter. That theme is the difficulty of administrative systems – any administrative system – coping with *change* in the economic and social fabric over which its net is cast. The more *rapid* the rate of change in the underlying socio-economic fabric is, the greater the difficulties of administrative systems 'to cope' become. All reforms of London government have experienced that difficulty, which has become more acute over time. Since it is expected that the rate of socio-economic change in the future will be ever faster than that in the past, it is no surprise that the conclusion of this chapter is, that an administrative approach based on an all-embracing big city government concept will never be able to offer a proper capacity 'to cope'.

It has been stated in the general context of European government reform – both in Western and in Eastern Europe – that administrative systems are naturally subject to obsolescence: once the political will has been assembled to implement reform, it is almost always the case that society and the economic system have moved on (Bennett, 1989a). Attempts to restrain the economy to the characteristics of the administrative system almost always lead to economic sluggishness and social disparity. The phenomenon of obsolescence increases in its impact with the rate of economic change. Because of this major constraint, the central argument, and conclusion, of this chapter is that solutions to the problem of the government of London can only be found in non-governmental as well as governmental structures – such as the use of the private sector, of market-driven public-private collaboration, and in local governments acting as 'competitive councils' which operate as quasi-market agencies.

London Government of a World City

'World cities' form that special group of cities that exercise leadership and perform key economic control functions in the global economy. They have been characterized by Peter Hall (1984) as the centres of major political power, often seats of national government and international agencies, and professional organizations. They are centres of trade, finance and government – these were their *raison d'être*, and it is these factors that have stimulated further population growth and sustenance of political and market control. World cities are dominated by their economic function. Their life-blood is their linkage into the global economic system.

For London the role of world city is built on three key elements. First is its key location in international financial markets. It is the dominant European centre of financial trading and one of the key international banking centres (see Chapter 5). As the major European financial centre it is linked by geography into a world role through its position between the time zones of North America and the Far East. Global trading of financial products has created a 24 hour service through switching work between centres with the daily opening and closing of the key gateways of London, New York and Tokyo/Hong Kong. These gateways then usually provide 'local' access – into European centres such as Frankfurt, Zurich, Paris or Amsterdam and into national financial centres, such as Glasgow, Edinburgh, Manchester or Leeds. This economic and financial role makes unique demands on London government, and particularly on its need for flexibility over the conflicting pressures produced by international economic demands and local needs.

The second key element is the role of London as the historic administrative centre of an empire which has now become the centre of the Commonwealth of Nations and of a wider network of international agencies. The demands this makes on prime sites, on airports, on communications and transport infrastructure places severe demands on London government to respond to an international framework.

Third is the element of London as a cultural and educational centre of world significance. London has more theatres than any other world city, as well as opera, ballet, museums and major educational institutions whose 'products', ranging from the Royal Ballet to the London School of Economics, have significance as world commodities with international 'brand recognition'. These products can only be consumed by being in London. The resultant demand on London government as a leisure, tourism and cultural capital again produces unique pressures on infrastructure, policing, cleansing services, and so forth.

The demands placed on world cities require minimalist local political intervention and maximum emphasis on response to the opportunities of the market for international products. This clearly creates tensions with local needs.

In terms of financial services, the main geographical base of government has been the City of London. This has retained a unique governmental structure. The electorate is primarily from the business community with a mayor who is largely a ceremonial head. Extensive delegation to officers is employed. Planning matters, although strongly oriented towards conservation of the buildings and character of the City of London, are dominated by maintaining the City as a world financial centre.

In recent years the strong concentration of financial services in the City of London has been diluted by restrictions on space to expand. This has led to a spread of financial services to fringe city and even to some more distant areas. The symbolic 'breakout' was the Broadgate development in Islington. Other breakouts have been into Victoria and King's Cross. However, the major development will be Canary Wharf, due for completion of phase one in 1992, which will increase financial services floor space by 25% of the present City of London. This breakout has not been easy for neighbouring boroughs to adjust to and has required major changes in land use plans and local political priorities. In the case of Canary Wharf, in Docklands, it has required special mechanisms to remove local political barriers to economic development – a central government Urban Development Corporation (which has assumed local planning powers) and the central government tax stimulus of an Enterprise Zone (giving relief from local property taxes and capital allowances against central corporation tax).

The sites of most of London's world services facilities are located in the inner boroughs – primarily in Kensington and Chelsea, Westminster, Camden, Islington, the City of London, Lambeth and Southwark. The world functions of London produce strong pressures on these central boroughs, but there are major spillover implications between London boroughs and between other local authorities in the rest of the metropolitan area. The spillovers are labour market flows, noise, congestion and other disbenefits, the airports in five areas around Heathrow, Gatwick, Stanstead, Luton and London City (in Docklands), and an array of financial burdens and benefits that are unequally distributed between areas.

Government for the National Capital

London is the seat of a national government and of most administrative functions. Although important counterweights are present in Scotland, Wales and Northern Ireland, and major decentralization of civil servants has occurred, London is still dominant in terms of number of administrative staff and governmental functions.

This has given to London a unique government structure whereby a number of functions that would be the responsibility of local authorities elsewhere are administered by central government itself. The Home Office, for example, has been responsible for policing in London (excluding the City of London) since 1829. Since the 1985 Local Government Act, central government direct powers have radically increased. Major examples are:

Land use planning   The Secretary of State for Environment is the strategic planning authority for London advised by the London Planning Advisory Committee (which is comprised of one representative from each borough).

Waste collection and disposal   This is delegated by the Secretary of State for Environment to the London Waste Regulation Authority for waste regulation and to a variety of joint borough authorities for waste collection.

Transport   The Secretary of State for Transport is directly responsible for many bridges, 208 miles of trunk and major roads, and has strong controls over a further 300 miles of road that are of strategic importance but are the boroughs' responsibility. The Secretary of State also has oversight of public transport through London Regional Transport and British Rail (for each of these functions see Hebbert and Travers, 1988).

London Government has a unique structure in which central government has strong powers to maintain those 'strategic' services which are required to satisfy national and international objectives. The present structure was strengthened after the abolition of the GLC and has thus moved strongly towards an emphasis on national and international objectives.

In addition to central government's role in local administration, it also appoints a number of agencies directly. The chief of these are:

- the Urban Development Corporation in London Docklands;
- the City Action Team focused mainly on Hackney, Islington and Lambeth;
- Greater London Arts;
- the London Division of the Historic Buildings and Monuments Commission;
- the London Council for Sport and Recreation;
- the Royal Parks.

London Government and the Metropolitan Area

The government of large cities is problematic in most parts of the globe. Academic ideas from the 1930s to the 1960s have tended to favour metropolitan government using an upper tier for metro-wide strategic planning, transport and land use control, and a more local tier for social and other policies (Stocks, 1939). The local government reforms in London, which created the London County Council in 1889 and the Greater London Council in 1965, were driven in major part by these arguments (Chapter 11). However, the abolition of the GLC in 1986 has both followed and led a rethinking of how metropolitan government in large cities should be organized in the future. The context of technological change, through its impact on the economy, has played a major part in leading to an evaluation of obsolescence for metro-government structures.

Metro-government has always had its problems. Its purpose has been to coordinate strategy over the metropolitan area. But what is the metro area? Definitions have usually followed a local labour market structure based on travel-to-work areas. A definition of 15% commuting has often been used to draw an outer boundary, following the definitions used for Standard Metropolitan Statistical Areas in the USA (see, for example, Berry, 1976). But a more rigorous criterion of 50% or 75% of commuting self-containment has usually been used to define the main area over which integrated government should be performed.

The arguments for a city region structure linking the populations to be served

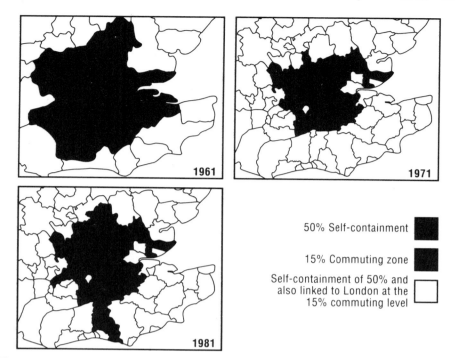

Figure 12.1 Self-containment of the London labour market and the 15% commuting zone 1961–81.

into units of size of at least 100 000–200 000 people played a large role in the conclusions of the Redcliffe-Maud Commission (Royal Commission on Local Government in England, 1969). That Commission was particularly influenced by the arguments for administrative economies of scale which they claimed were optimized over this size range. The evidence on which these views was based was somewhat suspect and contradictory. But even if the evidence had been reliable, the economics have been changed totally since 1965 by (i) the shifts in the size and shape of travel-to-work areas, and (ii) changes in the technology of service supply and its administration.

Figure 12.1 shows the metropolitan labour market for London based on census data for 1961, 1971 and 1981. What is immediately evident is the large span of commuting relationships over South East England and their change over time. The thoughts of earlier writers on local government were that ' . . . there is a growing expectation that local government areas should match real human communities' based on labour markets (Smart, 1974, 254); thus the Redcliffe-Maud Commission (1969, 3) held that inhabitants of a local government area ' . . . must share a common interest in their environment because it is where they live, work, shop and find their recreation'. These sentiments seem very idealized and unrealistic in the 1990s when we are now aware of the rapid changes in the dynamics of labour markets, of constant changes in investment strategies, and of immense cross-flows that make any 'sense' of self-containment in London labour markets unattainable. It would also be a modern sentiment that attempts to bind work, residence and recreation into unified administrative units. This represents, on the one hand, a hopelessly optimistic concept of integrated planning, and, on the other hand, an attempt to modify market dynamics by administrative action that creates needless barriers to the market and hence is incommensurate with the objectives of the economy as a whole. Indeed none of the four elements identified by the Redcliffe-Maud Commission as crucial principles for local government seem by current thinking to require a principle of labour market or other means of self-containment. The Royal Commission's elements were: (i) efficiency in performing tasks, (ii) attracting and holding the interest of citizens, (iii) developing strength to deal with national bodies, and (iv) adapting to the ' . . . process of change in the way people live, work, move, shop and enjoy themselves' (Royal Commission on Local Government in England, 1969, 1).

The spread of the commuter zone outside the London administrative area has always been an important issue. In 1866 it was estimated that 76 000 people commuted daily to the area that was to become the LCC, rising to 344 000 by 1911. By 1921 it had reached 843 000 and by 1961 it was 1 069 000 (Green,1950; Hall, 1964; and sources quoted in Smart, 1974). In the 1990s reverse and cross-commuting attained these orders of magnitude and, by 1981, the daily commuter flow into the former GLC area was one and a half million.

What the current period is suggesting, however, is not merely a need to respond to these labour market dynamics, but to rethink the paradigm of big city government as a whole. The problem of metropolitan governance is tied up with wider issues of what the role of local government as a whole should be in the economies of tomorrow.

London Government for Local Needs

London is a place in which almost seven million people live and three and a half million work within the 15 800 hectares of the 33 boroughs (including the City of London). Beyond this, 'London' is also a force in the South East England economy as a whole. Yet as well as catering for international and national demands, London government must also cater for local services to this population and its businesses.

The major organ of local government since 1986 has been the boroughs. But central government also plays a direct role in overseeing services, as it does in other parts of Britain, through its agencies which administer health, social security, employment and other functions. The London boroughs must now be conceived of as unitary authorities within the oversight powers of central government and its agencies.

The London boroughs are responsible for *education* in schools and further education colleges; *social services* for children and the elderly, to be increased in status with the 'Care in the Community' initiative; *environmental health* for noise, wastes, pest control, standards of food premises; *libraries*; *housing*; *arts, leisure and recreation*, which mainly means swimming pools, parks and sports areas; *refuse collection*; *land use planning*; and *roads and street lighting*. Of these services, education and housing are the two largest items of expenditure (see Hebbert and Travers, 1988).

Many boroughs have been in open conflict with central government over its role and responsibilities in London. This conflict has derived from three main sources. First has been the increasing socio-economic difficulties of many inner London boroughs. This has drawn the emphasis of local policies towards social and welfare issues, and has tended to make it difficult for boroughs to create a balance with the national and international needs of London. Rising social deprivation has also been seen as a political question since central government, which carries the major responsibility for social welfare, has been perceived to be somewhat aloof to the problems that are often all too visible to the boroughs. Indeed, central government has sought to adapt welfare support towards an ethic of self-help rather than dependency; and some boroughs have so far found this shift in emphasis difficult to adjust to. In fact, their political 'interest' orientation has often led them into open resistance.

Second has been the problem of a deep politicization of many boroughs intent on direct conflict with central government. A group of eight to ten London boroughs took on a strong 'new left' ideology during the period from approximately the mid-1970s up to 1987. In some boroughs strong elements still remain. The 'new left' is difficult to define. Gyford (1985, 18), for example, refers to it not as a 'single coherent ideology', but a 'syndrome of associated characteristics'. This 'syndrome' has *required* resistance to efforts to encourage management effectiveness and it has encouraged the politicization of administration. For example, Ken Livingstone, as chairman of the housing committee in Camden, 'reduced the number of meetings with the director of housing, and made contact with the third tier officers, who took day-to-day management decisions', and as leader of the GLC, he ' . . . wanted a new top administrative structure in which we could force early retirement on two-thirds of the chief officers, (so that the politically appointed) chairs of committees would have a predominant role in administration of departments' (Livingstone, 1987).

The third, and related, source of conflict has been management inefficiency. Central government has sought to gain economies and changed management style in local government throughout the nation by contracting out of services, by competitive tendering, and by closer checks through auditors. Some London boroughs have seen their policy instead as shaped towards 'creating' municipal employment.

The evidence for these trends in the mid-1980s is stunning. The Audit Commission (1986, 1987) drew attention to London's cycle of social and economic deprivation and management inefficiency arising from local politicization (see Figure 12.2). The Audit Commission argued that this was primarily a problem resulting from the objectives set by new left politicians of 'management by elected council members' with a proliferation of committees and subcommittees which were expensive in officers' time, in central overheads and in time delays. As a result it became difficult to recruit both officers and members, partly because of uncompetitive local government salaries for senior management officers, and partly because of continous political interference by members in officers' decisions. The Audit Commission found that turnover of senior officers in London boroughs such as Camden, Hackney, Haringey, Islington, Lewisham and Southwark was 40–69% in the three years 1983–86 and less than 20% of council members had more than four years experience. These relationships are further compromised by labour relations problems in which employees are reluctant to adapt to new work-practices. Elected members who are council employees or municipal trade unionists in inner London boroughs comprised between 33% and 50% of the Council members, with 61% in the former GLC (Walker, 1983, 1985). As a result they have found it difficult to distinguish their roles as politicians, employees and trade unionists: most attempts to increase efficiency resulted not in changes to work practices but in extra staff recruitment or salary regrading. High expenditure became inevitable.

The Audit Commission (1987) has revealed a wide variety of these deficiencies by comparing three groups of London boroughs: *group A* comprised eight deprived London boroughs with suspect management records; *group B* com-

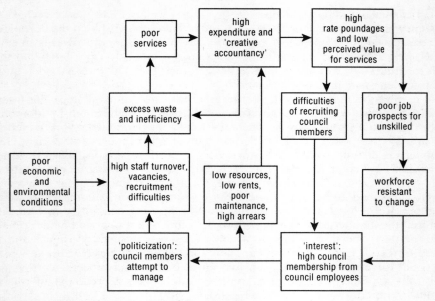

Figure 12.2   The cycle of decline in management skills

prised eight other London boroughs with similar problems of deprivation; and *group C* was comprised of eight of the most deprived metropolitan districts outside London. Their analysis demonstrates very high expenditure, levels of staffing and increases in staffing of group A compared with either groups B or C. These differences also apply to detailed items of housing, social services, etc. Group A authorities are usually 50% to 100% higher in both costs and personnel than authorities with comparable local socio-economic characteristics in London or elsewhere. The Commission concluded that ' . . . large parts of London appear set on precisely a course which will lead to financial and management breakdown . . . [which] are *not* due solely to social and economic factors over which they have no control' (Audit Commission, 1987, 4).

Analysis of the financial consequences of this policy demonstrate that between approximately 1979 and 1985 London boroughs had changed from rough parity in local tax rates to a situation in 1985 where 'new left' inner London had tax rates 36% higher than comparable areas, of which 26.4% (or 73% of the increase) was accounted for by 'new' expenditure (Bennett and Krebs, 1988). On business the burden of rates as a percentage of trading profits was up to three and a half times higher as a result (Bennett and Fearnehough, 1987; Bennett and Krebs, 1988). Up to 1985 this burden was supported by local taxes, but with central government's introduction of capping of local taxes in 1985, creative accountancy took over as the main mechanism.

The Audit Commission (1987) assessed the extent of the 'creative accountancy' in the same three groups of authorities discussed above. Creative accountancy involves three main mechanisms: use of balances accumulated from the past; borrowing to meet expenditure for maintenance that would normally be treated as revenue expenditure; and deferred purchase where borrowing is used to fund expenditure with interest rolled up into a first payment two or three years later. Group A authorities (all inner London 'new left') used these devices to produce a

Table 12.1 Percentage change in tax rates, 1974–85 due to different components of local authority budgets

	Total tax base	Need	Change in Rate Support Grant	Change in expenditure choices	Total
Inner London	−5.1	−10.1	−0.7	21.3	41.2
Outer London	−3.6	− 2.5	−4.6	10.6	10.1
London metropolitan region	−8.6	0.2	8.2	14.5	1.9
England and Wales	−7.9	−0.06	5.3	8.5	8.5

Source: simplified and recalculated from Bennett and Krebs, 1988; this table excludes City of London and Westminster; the boundary of London metropolitan region is that used in Figure 12.2 for 1981. Note all calculations are for lower tier authorities disaggregating county-level functions in proportion to local financial burden: see Bennett and Krebs, 1988, for details.

local debt burden two or three times as high as other areas. It now amounts to a debt of over £1000 per household for inner London, which will absorb 30% of the annual revenue expenditure for each year after 1987.

The outcome of these political developments has been strong differences between-boroughs in the financial balance of their resources and expenditure. Table 12.1 shows the consequence for the London metropolitan area. In this table budgetary components are translated into percentage changes in tax rates over the period 1974-85 deriving from a number of key factors. Over the period London gained greatly from increases in tax base. This allowed a reduction of tax rates. Inner London also gained greatly from a reduced need to spend as a result of population decline (see Chapter 9). Need was measured here by a complex index measuring over 70 service subcategories (see Bennett and Krebs, 1988). Although fixed costs do not decrease linearly with reductions in the number of clients, a reduction of up to 15% in real expenditure could have been possible from the combined effect of increased tax bases and the reduced need to spend. Instead, 'new' expenditure increased the budget by 21% in inner London, against 10–14% in other parts of the metropolitan area, and 8.5% nationally.

This pattern of change has interacted with the natural processes of social filtering in London. The difficulty has been that some local government action has tended to increase rather than diminish the social inequalities between areas. Bennett (1989b), for example, argues that there has emerged a London 'underclass' – the result of a vicious circle of positive feedback in which the social filtering of those able to migrate leaves behind an increasingly deprived population with progressively higher and higher dependence on public services and welfare support. Poor housing, low educational attainment, high youth unemployment and public service costs and inefficiency have each contributed to further social filtering. Those able to do so have had a natural tendency to search out alternative areas where the politicization of local government has taken on a different character. This process has been set amidst changes in the economic structure of the London economy as a whole which have made the dependent groups less and less able to participate because of inappropriate training or labour skills which the boroughs and the Inner London Education Authority have, until recently, done too little to address in their schools and colleges.

Major changes in thinking by most inner boroughs have taken place since 1987.

But the results of the decade of politicization from 1977 to 1987 have strongly confirmed the conflict of London government objectives between its local and its national and international roles.

London's Future Government?

The abolition of the Greater London Council in 1986 and the subsequent developments in London have given a shock to the debate about big city government. London had been a paradigmatic case, dating from the foundation of the LCC in 1889 and the expansion to 'strategic' functions with the GLC in 1965. Yet abolition has caused no pain and no seeming breakdown in services. In fact, in many people's perception, quite the reverse. Hall (1989, 184) sums up the debate: 'There is no God-given rule that local government is necessary', abolition is ' . . . as if government has proved its own assertion that the GLC was a monstrous and expensive irrelevancy' (Hall, 1989, 170).

How has such a paradigmatic change taken place? There seem to be six elements, each of which is briefly expanded on below.

Impossibility of strategic effectiveness

A major argument against the reinstitution of metropolitan government is the impossibility in practice of it ever being effective. In a market economy the economic signals are dispersed, changing and too complex for government fully to comprehend or to plan for. In a metropolitan area the size of London this problem is particularly evident. Similarly the political complexity of agreeing policy between 33 boroughs, and in some cases non-metropolitan counties and districts outside the GLC area, meant that it was impossible to take any strategic overview without strong command powers which are in conflict with the philosophy of either the market or of local democracy.

The impossibility of effectiveness is evident clearly in the history of the Greater London Development Plan, which was published in 1969, took four years to produce, three years to go through a public inquiry, and four years to get Secretary of State approval in a modified form. As a result, this attempt at *strategic* guidance instead produced a lacuna over a period of 11 years. The boroughs were bound to ignore it, and by the time it was agreed it was bound to be obsolescent.

The impossibility of effectiveness is further evident in the delicate balance of politics in the GLC area. This resulted in frequent changes in party control of the GLC and an inability to sustain a long-term policy in any strategic area. According to Hall (1989, 173) the GLC became increasingly ' . . . a political carnival rather than a serious strategic body for the management of the capital'.

Responsiveness of government to consumers

Due to the impossibility of strategic effectiveness, attention has turned to relating better to consumer demands, the services which London government is providing. This effort has been led by some London boroughs, notably Wandsworth, Westminster and Bromley. It has sought to withdraw local government from many service fields; and to redirect attention away from the concept of 'services', delivered to all in need as of right, instead to a concept of demand as expressed through preferences revealed through meaningful judgements stimulated by realization of the costs involved. This has stimulated the use of direct charging, vouchers, credits and administrative action. These methods are allowing greater emphasis to be placed on the smaller sized governmental units of the boroughs, an emphasis on 'market discipline' dominated by the unifying influence of central government towards a 'market' environment, and a greater possibility of variation to meet differing consumer needs through decentralization to the lowest tier of the London boroughs, as well as to supply companies and to agencies. This has been a key plank in the argument by the Audit Commission (1988) that local government must act as an opportunity seeker – as a 'competitive council' – to maximize the benefits of its activities for the people in its area. Whilst implementation is being led by some boroughs, it is clear that other boroughs will be forced into similar responsiveness by central government.

Managerial reform

A major plank of the central government's objectives has also been to improve local government management structures. The key objectives here have been the search for means to respond to the real demands of London's population, rather than to bureaucratic, political or 'interest' group decisions of what it is felt people should have. A key element has been that government employees should provide effective and cost-efficient services, rather than delivering services in conformity to labour union and previous work-practices (i.e. to assert consumer over producer interests). Key elements suggested by Foster (1988) are: (i) to improve governmental accounting practices, the way in which accounts are audited and publicized, and the way they are used within governmental decision-making; (ii) minimization of the managerial and financial links between units to make clear who is responsible for what – this requires major changes in 'corporate' government organization; (iii) to make responsible the head of each 'unit' to introduce responsibility and 'line-management' concepts; and (iv) to operate inter-unit transfers as closely as possible in relation to 'real' costs – this seeks to introduce a market discipline *within* the governmental organization.

Reinterpretations of representation

At the same time as innovation internal to government organization, major changes are also being urged in the area of the relations of government to the people it serves. Representation is the link in democracies between the government and its administrators, on the one hand, and to the individuals and businesses in its constituency, on the other hand. The issue of representation only arises in governmental systems of service delivery and finance. It is unnecessary where the mechanisms of production and finance are exchanges between people and businesses through market transactions. People in the market buy and sell what they choose and do not need the mediation of a politician. It is natural, therefore, that attempts to improve the effectiveness of government are stimulating pressures to 'privatize' services, contract them out, or otherwise modify governmental practice to replace 'political' or bureaucratic choice by customer choice.

Representation in London government is achieved through infrequent participation by voting for which turnouts are low (as they are in the rest of Britain). Local elections seek to offer agreements by the population on broad political agendas which legitimize a strategy for day-to-day administrative action. Even if this process works, it is not clear that it is the most efficient way of making decisions on detailed questions of how services are provided. Voting as a signal of market preferences is a very imperfect copy of real markets. If we accept that 'there is no God-given rule that local government is necessary' we open up a series of new questions of how best to introduce representation. Strategic planning can play little or no part in such approaches. Instead a means of direct purchase and sale of services gives the possibility of frequent signals of preference through fees or charges and market choices. These are more likely to lead to closer and better representation of customer interests without the mediation of politicians. This means radical reform to London government organization in the future. Although not all services can be charged for or removed from local 'political' decision, it is clear that there is considerable scope. The efforts of some local boroughs have proved the possibilities in the case of refuse collection, direct building and maintenance works, public housing, college-based training, and in some fields of personal social services. These developments are likely to bring London's services more in line with the more complex private, and public-private mixes present in almost all other world cities.

Finance and cost recovery

An OECD (1983) study argued that the finance of both infrastructure and service improvements by means other than through general budget (particularly tax) revenues may be more favourable to sustaining a balanced financial position for local government. This has stimulated major innovations towards 'cost recovery' rather than taxation; for example, through user charges, fees, licences, rentals and special assessments. At the same time major innovations in tax systems, both central and local, and the transfer system between levels of government, are also redirecting attention towards a cost-benefit relationship (i.e. towards benefit taxes such as the Community Charge or Poll Tax and charging methods). There are obviously a number of constraints on benefit taxes and user charges and other means of cost recovery. These arise from equity considerations, fluctuations in demand, and the relations between services financed by charges (or private sector provision) compared to those financed by public general revenues. However, it is becoming increasingly clear that within these constraints there is major scope for innovation. London government, like all governments, has too readily seen general revenue, particularly tax revenue, as the 'normal' financial source. It has indeed been the 'easy way' out of difficult decisions. Within the emerging paradigm, however, general sources are seen, as not only frequently less efficient, but also as frustrating the objective of achieving a proper relation of supply and demand. This has therefore necessarily led to greater use of mechanisms of exchange based on realistic pricing and other cost-recovery methods.

Shifting the boundary of government

Improving the responsiveness of services to customers, improving service delivery organization, managerial reform, reinterpretations of representation, and innovations in cost recovery can each be followed within almost any ideological framework. However, a major aim behind many recent emerging ideas has been a wider political agenda to shift back the boundary of the state, because it has been argued that less government, of itself, is a good thing. Three chief principles have been developed as a consequence (see the special 1988 issue of the journal *Economic Policy* on the 'Conservative Revolution'). The first principle is that the welfare gain to be derived from the pursuit of equity goals is less than the efficiency loss which government intervention produces. This is the so-called *price-of-fairness* principle. The second principle is the *fallibility of government*: that government will always do more harm than good when seeking to correct market failures because it is inherently less efficient than a market where incentives to good management and personnel performance are deemed always to assure greater efficiency. The third principle, and that which has been of most concern in inner London, is that government intervention has assured the *production and maintenance of dependency*. The provision of a service by government is seen as requiring moral judgements, on who receives or does not receive, by producer and 'interest' groups. Such a process inevitably leads to erosion of considerations to fit provision to means. At the same time, it has been argued, the ready acquiescence of politicians in the ambitious aims that have been drawn for services has led to expansions which of necessity undermine self-reliance and market alternatives. This fosters a view that government will solve all problems and hence fosters dependency: it becomes a 'producer state' or 'nanny state' (e.g. Pirie, 1988).

Conclusion

We do not have to go so far as accepting the need to shift the boundary of government to see that London's government in the future will need a greater flexibility and responsiveness to market demands than in the past – both inside and outside 'London' in its metropolitan area.

This chapter started out with the view that obsolescence of London government administrative areas was an inevitable consequence of rapid economic change. It has gone on to argue that seeking to institute concepts of self-containment into the government of London will impose rigidities to the economies of the market which will frustrate consumer choice and impose inefficiency on market allocation in the economy of both London and the country as a whole. The chapter has argued, on the one hand, that the demands on London government as an international and national city present strong limitations on the form of local government discretion possible within integrated global markets. On the other hand, the chapter has also argued that the pressures on government in the future will be to respond to customers and to be an adaptable market-like organization. Again this suggests the need for local government to exercise political and bureaucratic discretion within very tight economic disciplines.

None of the arguments presented suggest that the search for integrated London government is desirable or attainable. Rather the conclusion is that London local governments should be small in area, and flexible and responsive to the market stimuli on which their population and businesses depend. At the same time London boroughs will need to work more fully with each other, and with other actors in the private and voluntary sectors, and other public sector bodies. The result is likely to be a framework of *flexible aggregation* of local governments: inside and outside 'London' in the South East England metropolitan area; over different frameworks for different services; and over time. A number of new 'associations' and alliances between boroughs are likely to develop – indeed, many are already in place. The result will be a more complex pattern, a loss of strategic overview, but the possibility for a more responsive and efficient structure. It will also be a structure that more strongly corresponds to the flexible networks of inter-local authority co-operation that exists in most other world cities (e.g. in the US, Germany and Fránce). This development of flexible aggregation, association and flexible methods of service delivery are the inevitable consequences of London's local councils acting to maximize their individual opportunities and seeking the best suppliers of their services to the population and businesses in their areas. It marks the end of the paradigm of searching for a 'strategic' body for London.

13

London: An Uncertain Future
David R. Green and Keith Hoggart

London presents such a wealth of experience and variety of trends, themes and patterns, that commentaries on its future run the risk of being distracted by a multitude of significant, topical issues. This is a particular danger at this juncture in history, for, as Feldman and Jones (1989, 1) noted:

> In the events of the last decade, London has played not just a central but also a spectacular part. It is London which has been the site of the most visible triumph of a newly unfettered capitalism – the mushrooming of new financial services, the 'big bang', the transformation of the docklands, the dizzy rise of property values, the victory of the Murdoch empire over Fleet Street trade unionism . . . London is now the pre-eminent object of interest in everything from fashion to soap opera.

However, when considering London's future the temptation to incorporate every aspect of change needs to be resisted, otherwise dominant trends and potentialities can be obscured behind a mass of detail and nuances. Our intention in this postscript, therefore, is not to summarize the themes of earlier chapters in this book but to highlight the implications of the dominant trends they identified for London's future.

London 2000: Dimensions of Change

Predicting what lies ahead for London is extremely difficult. London is bound tightly into local, national and international economies (as well as companion political and social processes), so that the forces acting on the city are extremely complex. What is more, over time they appear to be becoming more volatile and unpredictable. This point is very evident if we examine the speculative writings of Peter Hall on London's future. Reflecting the seeming certainties of the 1960s, the 1963 edition of Hall's *London 2000* offered a predictive time span of almost 40 years. By contrast, and in obvious recognition of the rapidity of change since that time, when Hall presented his updated predictions in *London 2001* (Hall, 1989), the forward view had shrunk to little more than a decade. This difference is highly pertinent to our discussion, for it pinpoints a growing recognition that the forces which act on London are sufficiently powerful and volatile to generate considerable uncertainty over the capital's future.

Yet this does not mean that the forces most likely to prove central to change in the next decades are not known. In particular, of the major forces acting on London, three are likely to stand out as being of major significance: the role of

Britain in the world economy and the relative economic decline of the nation; the potentially disruptive forces of social polarization; and, the changing nature of political values.

With regard to the continuing decline of Britain's economic standing (Taylor, 1989), the peculiar position of London has to be recognized, for the roots of decline and some of its more flamboyant manifestations cannot be disentangled from the special role that powerful economic interests in London have had in directing trends within the British economy (e.g. Ingham, 1984). For Hall (1989, 51): 'In terms of employment, Britain is rapidly becoming a post-industrial economy, and in this respect London is merely an exaggerated example'. Yet, in our view, this understates the individuality of the city. Compared with else-where in the nation, manufacturing has long had a distinctive character in the capital (Chapter 2), and its continuing weakness has roots that go well beyond recent deindustrialization processes (Chapter 4). Hence, if we are to understand the prospects for London's future we have to be cognizant both of the peculiar standing of Britain in the world economy, and of the unique condition of London within Britain's economy, given the capital's overriding reliance on service indus-tries, and particularly the finance sector (Chapter 3).

The second major trend which is likely to have a significant role in deciding the future of the capital is social polarization. Once again there have been quite marked changes from the situation of two or three decades ago. The history of housing over the last 30 years provides a ready illustration of this. Contradicting trends so readily apparent at the time of *London 2000*, economic decline, coupled with shifts in political values, have produced sharp reversals in dominant themes By the late-1960s, economic collapse led to notable reductions in programmes for large-scale urban renewal (Dunleavy, 1981). High-rise construction was replaced by less costly programmes to upgrade the existing physical fabric of the city, which in turn opened the door for the social transformation of neighbour-hoods (e.g. Balchin, 1979), as they underwent gentrification. By the mid-1970s a further downturn in economic performance brought yet more pressure for lower public expenditure, which resulted in cut-backs across a broad spectrum of public housing programmes. Most recently, these economic exigencies have been intensified by ideologically-inspired government policies which have sought to invoke private sector solutions for housing problems. The result has been to emphasize market criteria for the allocation of housing investment, one conse-quence of which has been a steep rise in homelessness, such that today some 30 000 families in the capital (and many more single homeless people) are estimated to be without housing (London Research Centre, 1988). Transpar-ently, the cumulative effects of changes in London's housing situation over the last 30 years have been to increase the polarization of opportunities and living conditions for the capital's population. Yet polarization of housing opportunities is but one dimension of a wide pattern of increasing economic, political and social differentiation. In employment terms, for example, mismatches in worker skills and job availability have resulted in a much greater divergence in employment options and in income. The uneven geographical incidence of opportunities in sectors like employment and housing then make themselves manifest through the political framework of local government, resulting in strong distinctions between local policy priorities.

Adding further flavour to the city menu is the spice of changing values among political leaders. When Hall wrote *London 2000*, he did so against a backcloth of a post-war political consensus, in which the major parties perhaps disagreed over tactics and strategy, but shared many fundamental values concerning the future direction of British society. This was an era of 'One Nation' Toryism, wherein the rich were held to have an obligation to care for those less fortunate than themselves. How different from today, when the nation has experienced a decade-long experiment of new right efforts to unfetter *laissez-faire* capitalism. Even if all other aspects of society were stable, so sharp a diversion from the compromises of the 1960s would have severely limited the ability of writers in that decade to predict London's future.

To us, the combination of Britain's economic decline (and London's role within that process), of economic, political and social polarization, and of changing political values, is of crucial importance to understanding the future of the capital. The economic framework within which London operates has obvious relevance for it constrains the options that are open to policy-makers in seeking to bring about positive initiatives. Polarization likewise has a critical influence, since it helps determine the ability of institutions in the capital to organize collectively in order to meet the challenges of the future. Wrapped around and inter-woven with both of these forces is the impact of political values, for these provide a priority framework which affects whether the positive (or negative) impulses of economic change and social distinction will be smoothed or retarded. All three elements are key dimensions to any assessment of how the capital might change in the next 10–20 years.

London as a World City

In recent decades, internationalization of the global economy has allotted to those cities with financial control functions a crucial role in channelling international flows of capital (King, 1990). London currently retains its role as the largest currency trading centre in the world. With estimates recently placing its world share of currency dealing at 43.4%, it far exceeds both New York and Tokyo in this field. Hence, for the very obvious reason of profit, we can expect that at the very least British companies which are established in the capital will be keen to see London retain this position of dominance. Yet desires and actualities are not the same thing. Already the balance in world economic affairs is shifting. For instance, whereas in 1980 London's share of international bank lending was 27%, compared with a New York share of under 14% and a figure of 5% for Tokyo, by 1989 these percentages had shifted to 20, 10 and 21 respectively (*Guardian* 14 October 1989). Moreover, whilst London has gained in its share of financial futures and options trading, it has lost ground in each of bank lending, foreign exchange dealing, insurance and the holding of foreign funds.

In Europe also, London has rivals for its position of international dominance. Admittedly, as King (1990) argues, it does have an immense advantage over such centres as Frankfurt and Paris in that its natural language of business is the internationally favoured medium of English. Furthermore, it also has the advantage of relative cheapness. Despite high costs for office facilities relative to other European centres (Chapter 5), London has a notable comparative advantage in labour costs. Thus, according to the *Economist* (23 December 1989), the salaries of chief executives in firms with a turnover of more than £500 million average at £98 800 for Britain, compared with £318 400 for the United States, £211 800 for Japan and £107 800 for West Germany.

Yet what could counter the positive benefits of such advantages is the hesitant attitude of the present Conservative government towards European monetary union. Given Britain's late entry into the European Community, relatively few important Community institutions have been established in the capital. Without this tradition, and given the current government's often intransigent attitude towards our European partners, London may miss an opportunity to expand its role as an international focus for Europe. Perhaps this might only be a problem of short duration (assuming changes in governments in Britain and elsewhere), but already antagonism to Britain's spoiling tactics have helped European centres like Frankfurt to grow, aided in this case by a healthy, vibrant West German economy. With the opening up of Eastern Europe and the growing likelihood of increased integration between European Community countries and those of the former eastern bloc, the whole geographical focus of development may swing even further away from Britain, leaving the nation and its capital increasingly isolated on the western edge of an easterly expanding Europe.

Given the close relationship between London as a world city and political factors conditioning Britain's standing in the international community, it is clear that wider government policy is an important determinant of the city's future. At the very least, government policy and attitudes provide a context for decision-making in the private sector. This is readily seen in the Conservative government's response to the deregulation of the New York Stock Exchange in 1975. Admittedly, it was not until 1986 that the London Stock Exchange was deregulated. However, in forcing through deregulation, the Conservative government was attempting to counter the gains made by New York at the expense of London. This was not simply one further arm of the much touted new right desire to promote *laissez-faire*, but a deliberate tactic to counter a threat to London's primate world role in international finance. Unfortunately, for London as a whole, the Conservative government has been less willing to act in a positive manner to assist, maintain, or even enhance, London's position in other sectors of the economy. This is not a new departure for British governments, whether with regard to London or the nation as a whole, for the impetus behind favouring the financial sector has a much longer history than the life-span of one government (e.g. Ingham, 1984).

As seen in its assumption of the role of world leader in Eurodollar transactions a decade or more earlier, London has long been blessed with legal and institutional arrangements highly favourable to international financial dealings. As one example, the proportion of assets that foreign banks have to keep as security with the Bank of England is much lower than is usual in other nations, which has undoubtedly helped attract such institutions to London. Yet what must be set against these positive impulses is the visible appearance, again capable of being changed if the values of national political leaders are adjusted, of slippage in the high standards of conduct that the London financial markets used to pride themselves on. Characterized in much of the press as 'the sleaze factor', this slippage has been associated with an increase in the level of fraudulent, or at the very least highly questionable, practices. The Barlow Clowes affair in 1988 illustrates the problem, for it cast serious doubts on the Conservative government's ability (or willingness) to monitor investment practices and financial adviser integrity (as do the Guinness and County Natwest affairs). Furthermore, the government's failure to act on a damning report by the Department of Trade and Industry on irregularities in the take-over of Harrods and other House of Fraser establishments adds to the general sense that honesty is a less valued commodity in British business than was once the case.

Unless such views are counteracted they, along with the Conservative government's apparent antipathy towards Europe, could help weaken the future competitive position of London. Yet even if these two problems are eradicated, this might not alleviate the difficulties facing London's financial enterprises. As recent trading figures have revealed, it is possible that London is also in trouble on the economic trading front. Thus, in the last quarter of 1989, and for the first time since 1955, when data were first published on foreign currency exchanges in 'invisibles' (insurance premiums, profit repatriation, tourism, etc.), Britain fell into the red on its invisible balance of trade. What this recent decline shows is that the liberalization of financial markets has made the profitability of the City more insecure. In effect, the economic performance of the City, and through its domination of the capital's economy (Chapter 3), of London itself, has become more closely bound to the state of the world economy; an alignment that means that its future is more unpredictable and that its present economic foundations are more unreliable than ever before.

London 'Unhooked'

A further consequence of this international orientation to London's economy is that, as King (1990) has put it, London has become 'unhooked' from the rest of the British economy. In reality this polarization is between London and the expanding south-eastern economy and the rest of the nation, rather than London alone (e.g. Chapter 7 and Chapter 9). And it is also true that London has been 'unhooked' from the rest of the British economy since long before this century. Indeed, the City's prosperity never depended significantly on domestic manufacturing and has long relied on world trade, commerce and finance (Ingham, 1984), while the capital's manufacturing base has long operated from within a peculiarly local frame of reference (Chapter 2). Most especially in recent years, with British governments placing a higher priority on the sanctity of monetary targets, and concomitantly showing a lesser regard for advancing manufacturing, one result of this has been the growing intensity of the North–South divide within the nation (Lewis and Townsend, 1989). However, the fact that London's prosperity relies less on processes operating within the British economy and more on international factors relating to the global economy, has serious implications for the future of the capital. In particular, as this uncoupling has become more marked, the likelihood of either economic trends or governmental policies promoting economic expansion in *both* the London region and the rest of the nation at the same time has declined. Increasingly, the advancement of a London-based South-eastern economy runs the danger of being at the expense of the nation as a whole.

This feature is well illustrated in the manner in which policy priorities favourable to the financial sector over the past decade have placed short-term public expenditure targets ahead of the long-term requirements of economic revival. Cut-backs in funding for research and development are one aspect of this problem, with the postponement of capital investment in public transport, housing, education and health, to name a few central areas in what is a long list, weakening the infrastructural base which is required for an enhanced economy. What is essential to understand here is that the lack of investment in major items of infrastructure has now reached a critical point in some sectors. Indeed, at times the critical threshold has been crossed, with major disasters like the underground railway fire at King's Cross in 1987 and the Clapham Junction rail crash in 1988 bearing witness to the tragic effects that under-funded capital projects and insufficient maintenance spending can bring. Yet the lessons do not seem to have been learned, for the official response to the estimate that it will cost British Railways somewhere between £500 million and £1000 million to bring its facilities up to basic safety standards is that current passenger revenue must bear the brunt of this cost, at the same time as government grants for capital funding and subsidies for services are to be cut.

On a more local scale, polarization is also evident in the way that London is confronted with a peculiar combination of conflicting forces which arise from the competing demands of its role as an international city and from the pressures that emerge to satisfy its local population (Chapter 12). Take the issue of transport infrastructure to illustrate the point. As the city has become more of an arena for global financial transactions, conflict over scarce resources has led to the transport needs of the bulk of London residents being sacrificed in order to satisfy the preferences of internationally oriented office-based employment. Recently, this issue has been highlighted by the debate over new underground railway lines. It is widely accepted, even by the present Conservative government, that new lines are needed to relieve congestion in inner London. Yet, even though the government's own assessment favoured investment either in an east–west link between Liverpool Street and Paddington or else in a north–south link between Hackney and Chelsea, the scheme which has been placed at the top of the agenda is the extension of the Jubilee line from Green Park to Docklands. This facility is widely acknowledged as being capable of doing little to reduce inner London's general transport problems, but it will help make the giant office developments currently being built on the Isle of Dogs more attractive. Certainly, Docklands needs improved transport connections, for its light railway is far too slight to handle the number of commuters expected for developments like Canary Wharf (and the system is disadvantaged because its rolling stock cannot be used on the rest of the railway network). However, the key issue is not absolute need but opportunity costs. On this score we again see the impress of political values, for a major reason for the high priority given to the Jubilee line is simply that the developers, Olympia and York, who have a vested interest in Canary Wharf to the tune of £4 billion, offered to put up £400 million toward the cost of construction. Yet having obtained a government commitment to the Jubilee extension, Olympia and York are now reported to be backtracking from this funding commitment. That this seems to be linked to the company's engagement in negotiations over a £19 billion property deal in Tokyo is instructive, for it brings two points into perspective. First, it stresses that reliance on piecemeal solutions to communal problems is highly inappropriate, most especially when primacy is given to the availability of fluid and uncertain private sector funding. Secondly, it emphasizes how the future of London is tightly bound into a competitive struggle with other world cities, with major international corporations playing leading parts in selecting amongst a package of competing opportunities.

Polarization within London

In this struggle for a competitive edge the imagery that is projected concerning the economic prospects and the social conditions of cities is likely to have a crucial role. The importance of this point has been made clear in Hall's (1989) noting that the 'livability' of cities is becoming ever more influential in determining the success of world cities. On this score the future of London looks less bright. For one thing, the city's image, as with that of Britain as a whole, has not been enhanced by the periodic incidence of inner city riots during the 1980s (Clare, 1984; Greaves, 1984). Further tarnish has been added by the growing number and visibility of homeless people on its streets. These two cannot be divorced from one another, for they are linked to the growing bifurcation of economic conditions within the capital. Of course, London is a city that has long been characterized by sharp geographical contrasts in socio-economic circumstances (Jones, 1980; Green, 1985). However, compared with many other major cities, London retains an air of tranquillity and its few outbreaks of social tension have been on a lesser scale (Chapter 10).

Increasingly, however, with the collapse of manufacturing and the growing mismatch between traditional skills and the labour requirements of service sectors, many long-time residents of inner London boroughs have experienced reductions in employment opportunities at the same time as cuts in welfare programmes have occurred. High rates of inner city unemployment – reaching 20% in some areas, with even higher figures for black youths – contrast sharply with the experiences of the City. Indeed, in many ways places like Hackney and Tower Hamlets now have more in common with districts in northern cities like Moss Side and Toxteth than with the City of London (even some of London's suburban boroughs, like Barking and Havering, reveal a detachment from the rest of the capital in aspects of their employment structure; cf. Chapter 6).

The inflated salaries that were heralded by the 'big bang', along with the monetary bounties received by workers in international head offices and producer services, have ensured that some London residents take home an income sufficient to pursue a lifestyle largely unavailable elsewhere in the country. Of course, many aspects of that lifestyle are expensive, which restricts access to those of lesser means. In combination with the toll that the downward spiral of the British economy has taken on the value of sterling, this has meant that many of London's premier cultural attractions, like the theatre, opera and the symphonies, are inexpensive to foreign visitors, within comfortable reach of higher paid executives, but are too expensive for the bulk of residents to do more than attend at irregular intervals.

Cost also lies behind unequal access to housing. According to the Nationwide Anglia Building Society (1989), at a mean average price of £94 595, the cost of home purchase is 41% higher in London than in the nation as a whole, even though the average salaries of London purchasers are only 23% higher. What is more, Londoners obtain less for their money: fully 63% of purchased homes were built before 1945, compared with the national mean of 45%, and only 20% were detached or semi-detached properties (41% nationally). For a great majority of London residents, there is also a significant down-side to living in the capital in the form of the inconveniences of everyday life. High on the list is the cost and frustrations of using public transport. If current proposals are put into effect, Network South East will soon be the only commuter train service in Europe with no taxpayer subsidy (according to the Department of Transport's own assessment, in the year 2000 rail fares should be 46% higher in real terms while private motoring costs will increase by only 13%). This reflects the present Conservative government's antipathy toward public transport, but this in itself is merely an extreme manifestation of a long-term failure to invest in London's infrastructure. Yet decreased public transport services (Chapter 8), along with heightened passenger usage (notably for tourism in central areas), has placed tremendous strains on existing facilities. If these factors alone were not enough to reduce livability in the capital, many other indices of adverse social conditions, like crime rates, not only sit at the top of the national pile (and are accompanied by low clear-up rates), but are also subject to rapid rates of increase (between 1973 and 1984, for instance, recorded crimes rose by 77% and muggings by 400%; King, 1990, 147). Put all together, these conditions have produced one 1990 estimate that places the cost of living in London as 33.5% above the national average, which compares with an average middle manager's pay of 14.6% above the national norm, thereby giving London a net quality of living index of *minus* 18.9%, which places London as the least desirable place to live in Britain in terms of its 'quality of life' (*Guardian* 22 March 1990, 4).

What is more, this unpleasantness in London's living environment is socially discriminatory. Perhaps there are few Londoners who do not have to put up with the squeeze, turbulence, sparsity and delay of public transport operations, but it is not many. And when private transport modes have to be relied on, as they inevitably do, with a public transport system that, by international standards, puts itself to bed at an absurdly early time, which caters for little other than flows into and out of the centre and which barely recognizes new locations for basic facilities (like shopping; Chapter 7), the costs are unevenly felt. The higher costs of motor insurance in London takes larger proportionate slices from the incomes of the less well off (for a facility that is used less than is the case in the provinces), while uniform fines consequent upon falling foul of an inadequate car parking system take larger shares of disposable income from those earning less. Yet it is not simply in the way operational procedures are framed that polarization occurs, for the very institutional structures of London are permeated with forces favouring polarization. The abolition of the GLC has some part to play here (Bramley, 1984), with the abolition of the Inner London Education Authority in 1990 further widening inequities amongst boroughs. Yet these resource inequalities coincide with uneven incidences of service need (London Research Centre, 1986), quite pointedly seen in the geographical concentrations of children from deprived backgrounds and of immigrant parents (Chapter 10), both of which greatly increase schooling costs in certain inner London boroughs.

Of course, such unequal demands could be circumvented or at least ameliorated by the redistribution of governmental revenues, either within the capital or more appropriately at the national level. But even here forces of polarization seem strongly entrenched. One demonstration of this is seen in the enormous variation in charges announced for London boroughs in the first year of the poll tax (1990–91). At the time of writing, Lambeth has still to fix its poll tax figure, but it is expected that it will be over £600. At such, it will have the highest rate in England and Wales[1]. Accompanying Lambeth, London has the two other councils which have announced figures of more than £500, while, no doubt pursuing the mentality of second hand car sales, the capital has a further three councils that have announced figures no more than £2 under £500 (with almost all charges now formally published the highest non-London figure is £490.23 for Bristol). Yet look on the other side of the coin. Wandsworth has announced what is easily the lowest poll tax level at £148 per person (a figure that compares with the smallest rate for a non-London English council of £244.60 for Teesdale), one other London borough has come out with a figure below £200 (Westminster), and, out of only six councils in England and Wales with poll tax levels that make average household bills less than their 1989–90 property taxes, four are in London.

Yet it is not simply at the inter-borough level that discriminations occur. Evidence on this score is not plentiful, but a case can be made which points to boroughs unevenly favouring centres within their bounds (Chapter 11), while some have been seen to select (electorally inconsequential) neighbourhoods into which they 'dump' unwanted activities (see Glassberg, 1981, on Bromley). Both these effects have compounded discriminatory practices in the everyday operations of councils that produce significant variations in service provision across neighbourhoods (e.g. Tunley *et al.*, 1979).

What lies behind these divergences is a political environment that is characterized by more 'extremism' than the rest of the nation. The National Front, for example, as with its right-wing predecessors in earlier decades, has drawn its main body of support from London's East End (Husbands, 1982). Likewise, groups of the extreme left, such as Class War, find their most notable bastions of support within the capital. That this is the case cannot be divorced from the greater attention that these groups receive simply because they are in the capital, for this alone increases the likelihood that their actions will be reported, so popularizing their ideas and providing them with an enhanced sense of their own political efficacy. Such political bodies perhaps play a minor part in mainstream politics, but they reflect a broader tendency for the ideological impetus behind London politics to carry strong contradictory themes; one consequence of which is a tendency to indulge in petty, vindictive attacks on those people that political leaders do not favour (a recent example being the decision of the City of Westminster to count prisoners with homes in the borough as second home owners, so enabling the council to levy a double poll tax charge on them). Most certainly, London has not been characterized by the permanency of its radical politics, but when the national mood shifts towards radicalism its prominence is very apparent (the new left and the new right being recent manifestations of this). Combined with polarized social conditions, this means that the capital is likely to experience turmoil and conflict in its everyday existence, as well as in attempts to provide for the city's future. Irrespective of whether London has the administrative organization required for effective long-term response to emergent problems, within those structures London will always be liable to experience antagonism and upheaval on account of its polarized social and political conditions.

Organizing for Change

Most evident amongst the factors that will in the future handicap London's ability to respond to new opportunities and problems is its lack of a coordinating body that can integrate policy on a metropolitan-wide scale. Lest the reader recoils in anticipation that what will follow is a call for the reinstatement of the Greater London Council, rest assured. While we believe that Hall (1989, 170) is unduly critical in his assertion that the GLC '. . . did the things it was supposed to do badly or not at all, and it tried to do too many things that it should never have tried to do', we have to reiterate his point that the problems that brought the GLC into being have not gone away. Furthermore, as numerous commentators noted at the time of its abolition, many of the problems of the GLC were instituted by various national governments refusing to award this organization the functions and geographical scale it required to perform its tasks properly (e.g. Flynn *et al.*, 1985; Wheen, 1985). The story of the GLC's abolition in itself constitutes a sad little tale that reflects many of the problems that have beset London government from the time when the London County Council was formed in 1889. Perhaps history does not always repeat itself, but for the LCC and the GLC it certainly did. Both institutions were created by Conservative governments, both fell into the hands of their political opponents at the first election, and, perhaps not surprisingly, both experienced a Conservative backlash which sought to emasculate their role (with Conservative distrust already having been instrumental in their attaining lesser roles than others had wanted at the time these institutions were established). The 'truce' that existed between the Conservatives and these institutions was never an easy one. Always bubbling below the surface was the potential of a push for abolition (Young, 1975a). Yet, as recently as the late 1970s, when steps were taken by the Conservatives to evaluate the merits of abolition, the answers provided to the questions raised were that what was needed was a strengthening of the GLC's role, so that it could enhance the capacity of the capital to respond to emergent trends (Freeman, 1979). Perhaps London does not need a reinstated GLC, but it does need an organization that is capable of offering leadership for the capital.

Of course, in making this point we are conscious of a weakness in its logic. This is that it seems to imply that administrative structures for governance determine actual performance. In truth, as problems arising from social polarization visibly indicate, the relationship between these two is not straightforward. Within any established structure the range of behaviour options that are open to decision-makers is quite sufficient to give the appearance either of competent, precise, efficient management or of inexactitude, indecision and inappropriate action. Yet it is not feasible for us to predict what prevailing political attitudes will be in 10 years time. As we write this chapter the media are buzzing over the results of opinion poll surveys which reveal that if a parliamentary election was called today Mrs Thatcher would lose her own Finchley seat, and that Labour would win by a hefty majority. But we have seen similar polls in the past, most especially during the early 1980s, when war with Argentina helped bale the Conservatives out of the trough of their opinion poll decline (e.g. Norpoth, 1987). A similar unexpected event could occur again, which brings home to us that our assessments of the future must focus more on the (administrative as well as the socio-economic) frameworks within which the next decade or more will be played out, rather than the likely character of political leaders' attitudes.

Yet we must be aware that these frameworks are liable to change due to altered value stances amongst key policy-makers. The GLC provides a highly pertinent example of this, for amongst those who, in the late-1970s, strongly argued that the GLC should be saved and strengthened were a number of senior national Conservative politicians who would later take a leading part in its abolition (Wheen, 1985). What needs to be evaluated therefore is whether the present organizational structures for directing change in the capital (most obviously, but not exclusively, those of government) are capable of addressing likely constraints and opportunities that will arise over the next few decades. In our view they will not. There are certainly grounds for arguing that the somewhat lethargic procedures associated with an organization like the GLC are not well adapted to modern circumstances and that a system of government is required that is more readily able to adapt to emergent trends (Chapter 12). However, there is also an obvious need for the coordination of services and facilities, and for the active promotion of London's interests, over a much greater geographical area, and for many more functions than the current London boroughs cover. We cannot disguise the fact that the abolition of the GLC has left London without a voice which can seek coordination amongst its varied interests, which can look to the future over such issues as, what new investment will most advance its competitive position and which can lobby on the capital's behalf in the on-going competition for scarce national (and international) resources. Due to its lack of a metropolitan-wide or effective coordinating body for government, London now stands apart from the mainstream of major cities in the advanced economies (Norton, 1983). This, we believe, will severely handicap Britain's capital in what is a sharp competition over which world cities will be pre-eminent in 20 years time. Unlike New York or Paris, which have strong mayors who are national figures in their own right, London's government is divided between 33 'prime ministers' (Cousins, 1979). This encapsulates the problem. Certainly, as far as city-wide leadership is concerned, little can be expected of the boroughs. In addition to differences in their operating styles and in visions of their own purpose, inter-borough squabbling and strife are common (Chapter 11); as evinced in the existence of two, almost mutually exclusive, associations of London boroughs (the Conservative-dominated London Boroughs Association and the Labour-dominated Association of London Authorities).

Perhaps central government will take on the leadership mantle, for after all it did take over GLC functions with a metropolitan and beyond-metropolitan ambit (like public transport). At present, however, this seems so unlikely that it barely merits consideration. Capturing the seriousness with which the present Conservative government has reacted to issues of London government, O'Leary (1987a, 1987b) neatly summarized the process of the GLC's abolition as 'a farce'. This was an act of destruction for which, in marked contrast with the years of research and hearings that accompanied its establishment (Herbert Commission, 1960), was poorly and inconsistently explained and justified, hurriedly thought up and whose consequences appear to have been barely considered. All in all, abolition was put into practice with a rushed frenzy (Flynn *et al.*, 1985; Wheen, 1985).

Readers might rightly suggest that the unwillingness of the present Conservative government to offer leadership for London does not preclude such steps being taken by another government. This is true, but the point still has to be made that unless new and effective steps directly tackle problems of coordination and leadership, this will handicap the ability of other institutions within the capital to respond to emergent problems and prospects. What is more, the potential for such steps being effective cannot be considered to be good ones. For one thing, within the nation as a whole London politicians have long stood in a somewhat estranged position. As Feldman and Jones (1989, 1 and 3) noted for the Labour Party:

> Angry and exasperated provincial socialists round on London with talk of 'the London phenomenon' and treat it as a citadel of the 'loony left' ... [yet the] sectarianism, self-righteousness, and self-importance, as well as the energy and creativity, of metropolitan leftism have a venerable history ...

With such divisions between London and provincial party organizations the chances of breakdown between national political organizations and their London counterparts is intensified. And this phenomenon is not restricted to the Labour Party, for within Conservative ranks there has been a more ready appearance of new right councils in London, accompanied by events which have embarrassed and offended party representatives elsewhere (like the sale of Westminster's multi-million pound cemeteries for the princely sum of 15 pence). What is more, while London councils have not always worn a radical tinge, and indeed in some time periods have seemed to be characterized more by lethargy than political intent (e.g. Butterworth, 1966; Turner, 1978), what has distinguished London from other places is the durability of its radical urges and the intensity of its political party conflicts, both of which do make a visible appearance when the national climate is inclined towards political turmoil (e.g. Young, 1975b). Neither of these circumstances seem conducive to encouraging a new government to take special steps on London's behalf. Moreover, within the early 1990s at least, the scale of problems that a new national government will face, given decades of under-investment in basic infrastructure and highly unpopular cuts in standard services like education, health and housing, let alone the need to devote efforts to trying to improve both the long-term and the short-term performance of the British economy, will work against a new governmental arrangement for London being a high priority on any reform agenda.

But if national government is unlikely to act, will the private sector respond to the challenge? This is extremely unlikely. The very same reasons that deny London's boroughs the capacity to offer a strong, coordinated leadership for the capital are manifest in a wider range of contexts and with a sharper intensity in the private sector. For London, what is especially pertinent is the fact that the city's economy, as with that of the United Kingdom, is so open. Unlike local governments, which operate with a solid geographical fix, London's private sector institutions are not unavoidably tied to the capital: many are foreign-owned or have their strongest allegiance with another city; some are footloose, in that they are in London because of the present economic opportunities it offers but will be inclined to leave if the city loses its comparative advantages; and, then, there are those which are seemingly entrenched in London, but which might look for an alternative British site if the cost, the added inconveniences and the difficulties of employing and retaining quality staff in the capital worsen significantly (as already seen in the decentralization of some office employment). All in all, the prospects of the private sector providing long-term leadership aimed at the general well-being of the capital are slight. Where the prospects for positive private sector inputs are greater is in terms of seeking to maintain London's world city role.

The Future

We do not claim to have any special qualifications in the art of crystal ball gazing, but it does seem to us that the future of London will be conditioned by responses to a number of opportunities and constraints that currently confront it. In particular, we would identify the state of the British economy, the peculiar position of the capital within the national economy, the organization frameworks which ease or restrain effective responses to emergent trends, and the polarization of life within London, as important 'structural' dimensions to London life which will have critical impacts on future developments. For each of these our assessment is that the prospects for London are not good: the British economy looks set for yet more relative decline; the capital appears to be sharing less in the future of other regions than in the past; the organizational frameworks for effective communal responses are limpid; and patterns of social polarization seem to be leading to even more inequity, with its accompanying senses of grievance and implications for the livability of the city.

However, we have to recognize that structures do not determine actual events. Hence, if there is a change in governing political values, or indeed in their private sector counterparts, we could be looking at an altered frame of reference for London's future. This is critical, for trends in London life will be intensified or weakened by the value framework within which decisions are made. Since the late 1970s, for instance, national economic depression has increased the importance of right-wing control over the British government (Dunleavy, 1986). With an economy visibly in disarray *laissez-faire* approaches to economic 'management' were easier to justify, using the argument that state bureaucracy was largely responsible for handicapping the entrepreneurial drive of British capitalists. The combination of a decade of governments taking leading ideas from the new right and a forceful Prime Minister who has taken it on herself to seek to weaken opposition both inside and outside her political party has had significant implications for the future of London. In particular, it has intensified the degree to which London has become 'unhooked' from the rest of the British economy, by giving priority to financial interests over those of manufacturing (which has led to the south-eastern economy being much stronger than its provincial counterparts). In addition, through the priority given to profitability as a primary decision criteria, and through the deliberate attempt to weaken local government (most especially in the capital), London has been left with a woefully inadequate administrative capacity for responding to new challenges. At the same time, as displayed in the riots that have shaken the capital on various occasions in the 1980s, the preference for, indeed joy in, increased social polarization has made the capital a more uncomfortable place to live in; and has made serious inroads in the reputation of London and, more generally of Britain, as a place which is favoured with a desirable style of living.

All this suggests that London is moving toward a less advantageous place in the competition for ascendancy amongst world cities. But if the values of political leaders have intensified this trend, can a change in values ameliorate such adverse characteristics? Perhaps, but only with difficulty. For one thing, it will take time to undo much of the harm that has been done over the last decade; yet this must be achieved at a time when other cities have established a head of steam that has set them on the way to stealing an advantage over London. What is more, it is extremely unlikely that Britain's long-term decline in relative economic standing will be easily reversed. This is important for two reasons. First of all because this is an important element in London's continuing problems, as well as being an important handicap on new leaders implementing their preferred policies. Take as one example what could happen if a Labour government was elected to national office. Most likely, as happened to their predecessors in both 1964 and 1974, the new Labour administration would find itself under severe economic strain as capitalists revealed their uncertainties over the new government (or perhaps antagonism is more correct) by reducing investment and moving funds abroad. Strain on the economy in this sense could easily weaken the new administration's drive to change central features of the economy. But if the administration instead seeks to cultivate the approval of major private sector investors, then its policies could merely reinforce those trends which have already uncoupled London from the rest of the economy. This point is aligned with the second significant implication of the weakness of the British economy. For we would hold that if the economy as a whole is to experience improvement then this will entail a change in national priorities away from financial sectors toward other economic realms. But if this did occur, then, in the short-term at least, this could lead to substantial strain and disruption within the London economy. In a nutshell, London's future prospects do not look to be fortuitous; quite feasibly, heads London will lose, tails somewhere else will win.

Note

1   At the time of editing Lambeth's poll tax figure was £522, lower than expected but still amongst the highest in the country.

Bibliography

Abercrombie, P. (1945) *Greater London plan 1944*. London: HMSO.

Abu-Lughod, J. (1969) Migrant adjustment to city life: the Egyptian case. In Breese, G. (ed.) *The city in newly developing countries* (Englewood Cliffs, New Jersey: Prentice-Hall), 376–88.

Akers, J. (1987) 'Estate agents, urban managerialism and racial segregation in the London Borough of Enfield'. London: BA dissertation, King's College London.

Alexander, S. (1989) Becoming a woman in London in the 1920s and 1930s. In Feldman, D. and Jones, G. S. (eds.) *Metropolis. London* (London: Routledge), 245–71.

Arnold, W. (1986) *The historic hotels of London*. London: Thames and Hudson.

Ashford, D.E. (1986) *The emergence of the welfare states*. Oxford: Blackwell.

Ashford, S. (1987) Family matters. In Jowell, R., Witherspoon, S. and Brook, L. (eds.) *British social attitudes: 1987 report* (Aldershot: Gower), 121–52.

Association of London Authorities (1989) *Transport strategy for London*. London.

Audit Commission (1986) *Report and accounts, year ending 31 March 1986*. London: HMSO.

Audit Commission (1987) *The management of London's authorities: preventing the breakdown of services*. London: HMSO.

Audit Commission (1988) *The competitive council*. London: HMSO, Audit Commission Management Paper 1.

Baboolal, E. (1981) Black residential distributions in south London. In Jackson, P. and Smith, S.J. (eds.) *Social interaction and ethnic segregation* (London: Institute of British Geographers Special Publication 12), 59–79.

Bagguley, P. and Walby, S. (1989) Gender restructuring: a comparative analysis of five local labour markets. *Environment and Planning D: Society and Space* 7, 277–92.

Baines, D.E. (1981) The labour supply and the labour market 1860–1914. In Floud, R. and McCloskey, D. (eds.) *The economic history of Britain since 1700: volume two* (Cambridge: Cambridge University Press), 144–74.

Bakis, H. (1987) Telecommunications and the large firm: the case of IBM. In Hamilton, F.E.I. (ed.) *Industrial change in advanced economies* (London: Croom Helm), 130–60.

Balchin, P.N. (1979) *Housing improvement and social inequality: case study of an inner city*. Farnborough: Saxon House.

Ball, M. (1983)   *Housing policy and economic power*. London: Methuen.

Bank of England (1989)   London as an international financial centre. *Bank of England Quarterly Bulletin* November, 516–28.

Barlow, J. (1988)   Planning the London conversions boom: flat developers and planners in the London housing market. *The Planner* 75(1), 18–21.

Barratt Brown, A. (1934)   *The machine and the worker*. London: Weidenfeld and Nicolson.

Barton, N. (1962)   *The lost rivers of London*. London: Historical Publications.

Beck, G.M. (1951)   *A survey of British employment and unemployment 1927–45*. Oxford: University of Oxford Institute of Statistics.

Beier, A.L. and Finlay, R. (eds.) (1986)   *London 1500–1700*. Harlow: Longman.

Bennett, R.J. (ed.) (1989a)   *Territory and administration in Europe*. London: Frances Pinter.

Bennett, R.J. (1989b)   Resources and finances for the city. In Herbert, D.T. and Smith, D.M. (eds.) *Social problems in the city* (Oxford: Oxford University Press), 100–25 (second edition).

Bennett, R.J. (ed.) (1990)   *Decentralisation, local governments and markets: towards a post-welfare agenda*. Oxford: Oxford University Press.

Bennett, R.J. and Fearnehough, G. (1987)   The burden of the non-domestic rates on business. *Local Government Studies* 13(6), 23–36.

Bennett, R.J. and Krebs, G. (1987)   *Local business taxes in Britain and Germany*. Baden-Baden: Nomos.

Berry, B.J.L. (ed.) (1976)   *Urbanization and counterurbanization*. Beverly Hills: Sage.

Boal, F.W. (1970)   Segregation in west Belfast. *Area* 2, 45.

Booth, C. (1891)   *London labour and the London poor: first series – poverty*. London: Methuen (volume two).

Booth, C. (1902)   *London labour and the London poor: first series – poverty*. London: Methuen (volume five).

Booth, C. (1903)   *Life and labour of the poor in London: second series – industries*. London: Methuen (volume five).

Bowlby, S., Foord, J. and McDowell, L. (1986)   'For love not money: gender relations in local areas'. Newcastle upon Tyne: University of Newcastle upon Tyne Centre for Urban and Regional Development Studies Discussion Paper 76.

Bramley, G. (1984)   The distributional effects of abolishing the Greater London Council. *London Journal* 10, 46–54.

Branson, N. and Heinemann, M. (1973)   *Britain in the nineteen-thirties*. St Albans: Panther.

Bridge, M. (1989)   Which way congestion? *Transportation* 10(5), 187–90.

British Parliamentary Papers (Volume LXXXVII) (1852–53)   *Census for England and Wales for 1851*. London.

British Parliamentary Papers (1938)   *Royal commission on the distribution of the industrial population: minutes of evidence*. London.

British Parliamentary Papers (Volume IV) (1939–40)   *Royal commission on the distribution of the industrial population: report*. London.

Brosnan, P. and Wilkinson, F. (1987)   *Cheap labour: Britain's false economy*. London: Low Pay Unit.

Bruton, M. and Nicolson, D. (1987)   *Local planning in practice*. London: Hutchinson.

Buchanan, C. (1963)   *Traffic in towns*. London: HMSO

Buck, N., Gordon, I. and Young, K. (1986)   *The London employment problem*. Oxford: Clarendon.

Burgess, K. (1969) Technological change and the 1852 lock-out in the British engineering industry. *International Review of Social History* 14, 215–36.

Burnett, J. (1986) *A social history of housing 1815–1985*. London: Methuen.

Bush, J. (1984) *Behind the lines: East London labour 1914–19*. London: Merlin.

Butterworth, R. (1966) Islington Borough Council: some characteristics of single-party rule. *Politics* 1(1), 21–31.

Button, K.J. (1976) *Urban economics*. London: Macmillan.

Bythell, D. (1978) *The sweated trades*. London: Batsford.

Cannadine, D. (1984) The present and the past in the English industrial revolution 1880–1980. *Past and Present* 103, 131–72.

Carruthers, I. (1962) Service centres in Greater London. *Town Planning Review* 33, 5–31.

Cecchini, P. (1988) *The European challenge 1992*. Aldershot: Wildwood House.

Central Statistical Office (1989) *Regional trends 24*. London: HMSO.

Centre for Urban Studies (1964) *London: aspects of change*. London: Mac-Gibbon and Kee.

Champion, A.G. (1987a) An analysis of the recovery of London's population change rate. *Built Environment* 13, 193–211.

Champion, A.G. (1987b) Momentous revival in London's population. *Town and Country Planning* 56, 80–2.

Champion, A.G. and Congdon, P. (1988) Recent trends in Greater London's population. *Population Trends* 53, 7–17.

Champion, A.G. *et al.* (1987) *Changing places: Britain's demographic, economic and social complexion*. London: Edward Arnold.

Chance, J. (1987) The Irish in London. In Jackson, P. (ed.) *Race and Racism* (London: Academic), 142–60.

Chartered Institute of Public Finance and Accountancy (1989) *Local government comparative statistics*. London.

Chisholm, C. (1938) *Marketing survey of the United Kingdom and census of purchasing power distribution*. London: Business Publications.

Clare, J. (1984) Eyewitness in Brixton. In Benyon, J. (ed.) *Scarman and after* (Oxford: Pergamon), 46–53.

Clark, C. (1951) Urban population densities. *Journal of the Royal Statistical Society,* A14, 490–96.

Clayton, R. (ed.) (1964) *The geography of Greater London*. London: George Philip and Son.

Clout, H.D. and Wood, P.A. (eds.) (1986) *London: problems of change*. London: Longman.

Collins, M.F. and Pharoah, T.M. (1974) *Transport organisation in a great city: the case of London*. London: Allen and Unwin.

Commission for Racial Equality (1984) *Race and council housing in Hackney*. London.

Congdon, P. (1983) *Map profile of Greater London 1981*. London: GLC Statistical Series 23.

Congdon, P. (1984) *Social structure in the London boroughs: evidence from the 1981 census and changes in 1971*. London: GLC Statistical Series 28.

Congdon, P. (1989) An analysis of population and social change in London wards in the 1980s. *Transactions of the Institute of British Geographers* 14, 478–91.

Congdon, P. and Champion, A.G. (1989) Trend and structure in London's migration and their relation to employment and housing markets. In Congdon, P. and Batey, P. (eds.) *Advances in regional demography* (London: Belhaven), 180–204.

Coote, A. and Campbell, B. (1982)   *Sweet freedom*. Oxford: Blackwell.

Coppock, J.T. and Prince, H.C. (eds.) (1964)   *Greater London*. London: Faber and Faber.

Corporation of London (1988)   *City of London local plan: second monitoring report*. London.

Corporation of London (1989)   *City of London local plan*. London: Corporation of London Department of Planning.

Courtney, N. (1987)   *The luxury shopping guide to London*. London: Weidenfeld and Nicolson.

Cousins, P.F. (1979)   Council leaders – London's 33 prime ministers. *Local Government Studies* 5(5), 35–46.

Cousins, P.F. (1988)   PGOs in London. In Hood, C. and Schuppert, G. F. (eds.) *Delivering public services in Western Europe: sharing Western European experience of Para-government organization*. (London: Sage), 155–165.

Crafts, N.F.R. (1987)   Long term unemployment in Britain in the 1930s. *Economic History Review* 40, 418–32.

Crompton, R. and Mann, M. (eds.) (1986)   *Gender and stratification*. Cambridge: Polity.

Cronin, J. (1979)   *Industrial conflict in modern Britain*. London: Croom Helm.

Cronin, J. (1984)   *Labour and society in Britain 1918–79*. London: Batsford.

Daly, M. and Atkinson, E. (1940)   A regional analysis of strikes 1921–36. *Sociological Review* 32, 216–23.

Dasgupta, M. *et al.* (1985) 'Factors affecting mode of choice for the work journey'. Crowthorne: Transport and Road Research Laboratory Research Report 38.

Dasgupta, M. *et al.* (1989)   'Journey to work trends in British cities'. Crowthorne: unpublished report to the Transport and Road Research Laboratory.

Davies, K. and Sparks, L. (1989)   The development of superstore retailing in Great Britain 1960–86. *Transactions of the Institute of British Geographers* 14, 74–89.

Davis, J. (1988)   *Reforming London: the London government problem 1855–1900*. Oxford: Clarendon.

Deakin, N. and Cohen, B. (1970)   Dispersal and choice: a study of West Indians in London. *Environment and Planning* A2, 193–210.

Dearlove, J. (1973)   *The politics of policy in local government: the making and maintenance of public policy in the Royal Borough of Kensington and Chelsea*. Cambridge: Cambridge University Press.

Denton, P. (1988)   *Betjeman's London*. London: John Murray.

Department of the Environment (1977)   *Inner London: policies for dispersal and balance*. London: HMSO (final report for the Lambeth Inner Area Study).

Department of the Environment (1977a)   *Change and decay*. London: HMSO (final report for the Liverpool Inner Area Study).

Department of the Environment (1977b)   *Unequal City*. London: HMSO (final report for the Birmingham Inner Area Study).

Department of the Environment (1989)   *Draft strategic planning guidance for London*. London.

Department of the Environment (1989a)   *Strategic planning guidance for London*. London.

Department of the Environment/Roger Tym and Partners: *Monitoring enterprise zones*. London: Roger Tym and Partners (various years).

Department of Transport (1986a)   *Crime on the Underground*. London: HMSO.

Department of Transport (1986b)   *East London assessment study: stage one report – transport problem identification*. London.

Department of Transport (1988) *Transport statistics for London.* London: Department of Transport Statistics Bulletin (88) 51.

Department of Transport (1989a) *Statement on transport in London: the Secretary of State's approach towards the operation and development of London's transport systems.* London.

Department of Transport (1989b) *Central London rail study.* London.

Department of Transport (1989c) *Roads for prosperity.* London: HMSO.

Dex, S. (1988) Gender and the labour market. In Gaillie, D. (ed). *Employment in Britain.* (Oxford: Blackwell), 281–309.

Diamond, I. and Clarke, S. (1989) Demographic patterns among Britain's ethnic groups. In Joshi, H. (ed.) *The changing population in Britain.* (Oxford: Blackwell), 177–98.

Doherty, J.M. (1973) 'Immigrants in London: a study of the relationship between spatial structure and social structure'. London: PhD thesis London School of Economics.

Dolphin, P. *et al.* (1981). *The London region: an annotated geographical bibliography.* London: Mansell

Donnison, D. and Eversley, D. (eds.) (1973) *London: urban patterns, problems and policies.* London: Heinemann.

Drivers Jonas (1989) *Croydon shopping study.* Croydon: London Borough of Croydon.

Duncan, S.S. (1990) 'The geography of gender divisions of labour in London'. Brighton: University of Sussex Urban and Regional Studies Working Paper 73.

Duncan, S.S. (forthcoming) 'The geography of gender divisions of labour in Britain'. Brighton: University of Sussex Urban and Regional Studies Working Paper.

Duncan, S.S. and Savage, M. (1989) Space, scale and locality. *Antipode* 31, 179–206.

Dunleavy, P.J. (1981) *The politics of mass housing in Britain 1945–75.* Oxford: Clarendon.

Dunleavy, P.J. (1986) The growth of sectoral cleavages and the stabilization of state expenditures. *Environment and Planning D: Society and Space* 4, 129–44.

Dunning, J.H. (1958) *American investment in British manufacturing industry.* London: Allen and Unwin.

Dyos, H.J. (1961) *Victorian suburb: a study of the growth of Camberwell.* Leicester: Leicester University Press.

Dyos, H.J. (1982) A guide to the streets of London. In Cannadine, D. and Reeder, D. (eds.) *Exploring the urban past* (Cambridge: Cambridge University Press), 190–201.

Dyos, H.J. and Aldcroft, D.H. (1974) *British transport: an economic survey from the seventeenth century to the twentieth.* Harmondsworth: Penguin.

Elkin, S.L. (1974) *Politics and land use planning: the London experience.* Cambridge: Cambridge University Press.

Essberger, S. (1987) *Monopoly London: the Monopoly player's tour of London.* London: Chameleon.

Evans, A.W. (1973) The location of headquarters of industrial companies. *Urban Studies* 10, 387–95.

Feagin, J.R. and Smith, M.P. (1987) Cities and the new international division of labour. In Smith, M.P. and Feagin, J.R. (eds.) *The capitalist city.* (Oxford: Blackwell), 3–34.

Feldman, D. and Jones, G.S. (1989)   Introduction. In Feldman, D. and Jones, G.S. (eds.) *Metropolis. London* (London: Routledge), 1–7.

Field, B. (1982)   'The evolution of London's planning system and the changing role of the boroughs'. London: South Bank Polytechnic Department of Town Planning Occasional Paper 2/82.

Field, B. (1983)   Local plans and local planning in Greater London: a review. *Town Planning Review* 54, 24–40.

Field, B. (1984)   Theory in practice: the anatomy of a borough plan. *Planning Outlook* 27(2), 68–78.

Fielding, A.J. (1989)   Inter-regional migration and social change: a study of South East England based upon data from the Longitudinal Study. *Transactions of the Institute of British Geographers* 14, 24–36.

Finlay, A. (1981)   *Population and metropolis: the demography of London 1580–1650.* Cambridge: Cambridge University Press.

Finlay, I.F. (1986)   *London in old picture postcards.* Zaltbommel: European Library.

Fisher, F.J. (1948)   The development of London as a centre of conspicuous consumption in the 16th and 17th centuries. *Transactions of the Royal Historical Society* 30, 37–50.

Flynn, N. *et al.* (1985)   *Abolition or reform? the GLC and the metropolitan county councils.* London: Allen and Unwin.

Foley, D.L. (1963)   *Controlling London's growth: planning the Great Wen 1940–1960.* Berkeley: University of California Press.

Forsham, A. and Bergstrom, T. (1986)   *The open spaces of London.* London: Allison and Busby.

Foster, C.D. (1988)   Accountability in the development of policy for local taxation of people and business. *Regional Studies* 22, 13–8.

Fox, C. (1987)   *Londoners.* London: Thames and Hudson.

Freeman, R. (1979)   The Marshall Plan for London's government. *London Journal* 5, 160–175.

Frost, M.E. and Spence, N.A. (1984)   The changing structure and distribution of the British workforce. *Progress in Planning* 21, 67–147.

Frost, M.E. and Spence, N.A. (1991)   Understanding employment change in central London. *Geographical Journal* 157(1).

Garside, P.L. (1983)   Inter-governmental relations and housing policy in London 1919–1970, with special reference to the density and location of council housing. *London Journal* 9, 39–57.

Gershunny, J. (1983)   *Social innovation and the division of labour.* Oxford: Oxford University Press.

Gibbs, A. (1987)   Retail innovation and planning. *Progress in Planning* 27, 7–67.

Giddens, A. (1984)   *The constitution of society: outline of the theory of structuration.* Berkeley: University of California Press.

Gilje, E. (1983)   *Demographic review of Greater London.* London: GLC Statistical Series 20.

Gillespie, J. (1989)   Poplarism and proletarianism: unemployment and Labour politics in London 1918–34. In Feldman, D. and Jones, G.S. (eds.) *Metropolis. London* (London: Routledge), 163–88.

Glading, P. (1934)   *How the Bedaux system works.* London: Labour Research Department.

Glass, R. (1960)   *Newcomers: the West Indians in London.* London: Allen and Unwin.

Glass, R. (1964) Introduction. In Centre for Urban Studies *London: aspects of change* (London: MacGibbon and Kee), xiii–xiii.

Glassberg, A. (1973) The linkage between urban policy outputs and voting behaviour: New York and London. *British Journal of Political Science* 3, 341–361.

Glassberg, A. (1981) *Representation and urban community.* London: Macmillan.

Goddard, J.B. (1967) The internal structure of London's central area. In *Urban Core and Inner City* (Leiden: Brill), 118–40.

Goddard, J.B. (1968) Multivariate analysis of office location patterns in the city centre: a London example. *Regional Studies* 2, 69–85.

Goodman, A. (1988) *Gilbert and Sullivan's London.* Tunbridge Wells: Spellmount Ltd.

Gottman, J. (1964) *Megalopolis.* Cambridge, Mass.: MIT Press.

Gottman, J. (ed.) (1980) *Centre and periphery: spatial variations in politics.* Beverly Hills: Sage.

Greater London Council (1969) *Greater London development plan statement.* London.

Greater London Council (1976) *Greater London development plan.* London.

Greater London Council (1976a) *Greater London development plan: written statement* (as approved by the Secretary of State for the Environment on 9 July 1976). London.

Greater London Council (1984) *The Greater London development plan, as proposed to be altered by the Greater London Council.* London.

Greater London Council (1985a) *Greater London transportation survey 1981.* London.

Greater London Council (1985b) *London industrial strategy.* London.

Greater London Council (1985c) *Women on the move: GLC survey on women and transport.* London: GLC London Strategic Policy Unit.

Greater London Council and the Department of the Environment (1974) *London rail study.* London.

Greaves, G. (1984) The Brixton disorders. In Benyon, J. (ed) *Scarman and after* (Oxford: Pergamon), 63–72.

Green, D.R. (1985) A map for Mayhew's London: the geography of poverty in the mid-nineteenth century. *London Journal* 11, 115–26.

Green, D.R. and Parton, A. (1990) Slums and slum life in Victorian England: London and Birmingham at mid-century. In Gaskell, M. (ed.) *The slum* (Leicester: Leicester University Press), 17–91.

Green, F.H.W. (1950) Urban hinterlands in England and Wales: an analysis of bus services. *Geographical Journal* 116, 64–88.

Greve, J. *et al.* (1971) *Homelessness in London.* Edinburgh: Scottish Academic Press.

Gyford, J. (1985) *The politics of local socialism.* London: Allen and Unwin.

Halford, S. (1988) Women's initiatives in local government: where do they come from and where are they going? *Policy and Politics* 16, 251–60.

Halford, S. (1989) Spatial divisions and women's initiatives in British local government. *Geoforum* 20, 161–74.

Hall, J.M. (1976) *London: metropolis and region.* Oxford: Oxford University Press.

Hall, P.G. (1962) *The industries of London since 1861.* London: Hutchinson.

Hall, P.G. (1964) The development of communications. In Coppock, J.T. and Prince, H.C. (eds.) *Greater London* (London: Faber and Faber), 52–79.

240   *Bibliography*

Hall, P.G. (1966)   *The world cities*. London: Weidenfeld and Nicolson.
Hall, P.G. (1969)   *London 2000*. London: Faber and Faber (second edition).
Hall, P.G. (1989)   *London 2001*. London: Unwin Hyman.
Hall, P.G. and Preston, P. (1988)   *The carrier wave: new information technology and the geography of innovation 1846–2003*. London: Unwin Hyman.
Hall, P.G. *et al.* (1987)   *Western sunrise: the genesis and growth of Britain's major high tech corridor*. London: Allen and Unwin.
Hall, S. *et al.* (1978)   *Policing the crisis: mugging, the state and law and order*. London: Macmillan.
Hamilton, F.E.I. (1976)   Multinational enterprise and the European Community. *Tijdschrift voor Economische en Sociale Geografie* 67, 258–78.
Hamilton, F.E.I. (1986)   Industrial organization and regional labour markets. In Fischer, M.M. and Nijkamp, P. (eds.) *Regional labour markets: analytical contributions and cross-national comparisons* (Amsterdam: North Holland), 289–312.
Hamnett, C.R. (1976)   Social change and social segregation in inner London 1961–71. *Urban Studies* 13, 261–72.
Hamnett, C.R. (1986)   The changing socio-economic structure of London and the South East 1961–81. *Regional Studies* 20, 391–406.
Hamnett, C.R. and Randolph, W. (1982)   How far will London's population fall? *London Journal* 8(1), 95–100.
Hart, D. (1976)   *Strategic planning in London: the rise and fall of the primary road network*. Oxford: Pergamon.
Harvey, D.W. (1985)   *The urbanization of capital*. Oxford: Blackwell.
Head, A. (1957)   *Sign boards of old London shops*. London: Portman (reissued 1988).
Healey, P. (1983)   *Local plans in British land use planning*. Oxford: Pergamon.
Hebbert, M. and Travers, T. (eds.) (1988)   *The London government handbook*. London: Cassell.
Her Majesty's Stationery Office (1965)   *Report of the Committee on Housing in Greater London*. (chair: Sir Milner Holland) London.
Her Majesty's Stationery Office (1973)   *Greater London development plan: report of the panel of inquiry*. (chair: Sir Frank Layfield) London.
Her Majesty's Stationery Office (1981)   *The Brixton disorders 10–12 April 1981*. (chair: Lord Scarman) London.
Her Majesty's Stationery Office (1987)   *Bangladeshi in Britain*. London.
Her Majesty's Stationery Office (1989)   *Domestic violence: an overview of the literature*. London.
Herbert Commission (1960)   *Royal commission on local government in Greater London 1957–60: report*. London: HMSO (cmnd 1164).
Hewitt, P. and Mattinson, O. (1988)   'Women's votes: the key to winning'. London: Fabian Research Series 353.
Hilgert, F. (1945)   *Industrialization and foreign trade*. Geneva: League of Nations.
Hillier Parker (1987)   *ICHP rent index 21*. London.
Hiller Parker (1989)   *Retail parks*. London.
Hillman, M., Henderson I. and Whalley, A. (1976)   *Transport realities and planning policy*. London: Political and Economic Planning.
Hinton, J. (1983)   *Labour and socialism: a history of the British labour movement 1867–1974*. Brighton: Harvester.
Hiro, D. (1971)   *Black British white British*. London: Eyre and Spottiswoode.
Hobsbawm, E. (1964)   The nineteenth-century London labour market. In Centre

for Urban Studies *London: aspects of change* (London: MacGibbon and Kee), 3–28.

Hobsbawm, E. (1984) *Worlds of labour*. London: Weidenfeld and Nicolson.

Hodgson, M. (1984) *Women living in London*. London: GLC Statistical Series 34.

Holmes, C. (1982) The impact of immigration on British society, 1870–1980. In Barker, T.C. and Drake, M. (eds.) *Population and society in Britain 1850–1980* (London: Batsford), 172–202.

Howard, E.B. and Davies, R.L. (1988) *Change in the retail environment*. Harlow: Longman.

Howarth, E. and Wilson, M. (1907) *West Ham: a study of social and industrial problems*. London: Dent.

Howells, J. and Green, A.E. (1986) Location, technology and industrial organisation in U.K. services. *Progress in Planning* 26(2), 83–184.

Humphries, J. (1983) The 'emancipation' of women in the 1970s and 1980s: from the latent to the floating. *Capital and Class* 20, 6–28.

Husbands, C.T. (1982) East End racism 1900–1980: geographical continuities in vigilantist and extreme right-wing political behaviour. *London Journal* 8, 3–26.

Husbands, C.T. (1985) Attitudes to local government in London: evidence from opinion surveys and the GLC by-election of 20 September 1984. *London Journal* 11, 59–74.

Hutt, A. (1937) *The condition of the working class in Britain*. London: Martin Lawrence.

Huws, U. (1984) *The new homeworkers*. London: Low Pay Unit.

Ingham, G. (1984) *Capitalism divided? the City and industry in British social development*. London: Macmillan.

Jackson, B. and Jackson, S. (1979) *Childminder: a study in action research*. London: Routledge and Kegan Paul.

Jackson, P. (1986) Ethnic and social conflict. In Clout, H.D. and Wood, P.A. (eds.) *London: problems of change* (London: Longman), 152–59.

Jackson, P. (1987) The idea of race and the geography of racism. In Jackson, P. (ed.) *Race and racism* (London: Academic), 3–21.

Jones Lang Wootton (1988) *Central London offices research*. London.

Jones Lang Wootton (1989) *The city office review 1984–88*. London.

Jones Lang Wootton (1989a) *Central London office research*. London.

Jones, E. (1980) London in the early seventeenth century: an ecological approach. *London Journal* 6, 123–33.

Jones, E. (1985) The Welsh in London in the nineteenth century. *Cambria* 12, 149–69.

Jones, E. and Eyles, J. (1977) *Introduction to social geography*. Oxford: Oxford University Press.

Jones, G.S. (1971) *Outcast London: a study in the relationship between classes in Victorian society*. Oxford: Clarendon.

Jones, G. (1988) Foreign multinationals and British industry before 1945. *Economic History Review* 41, 429–53.

Keating, P.J. (1971) *The working classes in Victorian fiction*. London: Routledge and Kegan Paul.

Kelly, L. (1988) *Surviving sexual violence*. Cambridge: Polity.

King, A.D. (1990) *Global cities: post-imperialism and the internationalization of London*. London: Routledge.

Laite, J. and Halfpenny, P. (1987) Employment, unemployment and the domestic division of labour. In Fryer, D. and Ullah, P. (eds.) *Unemployed people* (Milton Keynes: Open University Press), 194–216.

Lane, B. (1988)   *The murder club guide to London*. London: Harrap.

Lash, S. and Urry, J. (1987)   *The end of organized capitalism*. Cambridge: Polity.

Lee Donaldson Associates (1986)   *Superstore appeals review 1986*. London.

Lee Donaldson Associates (1988)   *Superstore appeals review 1988*. London.

Lee, C.H. (1986)   *The British economy since 1700*. Cambridge: Cambridge University Press.

Lee, T.R. (1977)   *Race and residence: the concentration and dispersal of immigrants in London*. Oxford: Oxford University Press.

Lejeune, A. (1979)   *The gentlemen's clubs of London*. London: Bracken Books.

Lewis, J. and Townsend, A. (eds.) (1989)   *The North-South divide*. London: Paul Chapman.

Leyshon, A. and Thrift, N.J. (1989)   South goes North? the rise of the British provincial financial centre. In Lewis, J. and Townsend, A. (eds.) *The North-South divide* (London: Paul Chapman), 114–56.

Linge, G.J.R. and Hamilton, F.E.I. (1981)   International industrial systems. In Hamilton, F.E.I. and Linge, G.J.R. (eds.) *Spatial analysis, industry and the industrial environment volume two: international industrial systems* (Chichester: Wiley), 1–114.

Lipman, V.D. (1954)   *Social history of the Jews in London 1850–1914*. London: Routledge and Kegan Paul.

Lipman, V.D. (1964)   Social topography of a London congregation: the Bayswater synagogue 1863–1963. *Jewish Journal of Sociology* 6, 69–74.

Littler, C. (1983)   Deskilling and changing structures of control. In Wood, S. (ed.) *The degradation of work?* (London: Hutchinson), 122–45.

Livingstone, K. (1987)   *If voting changed anything they'd abolish it*. London: Collins.

Llewellyn Smith, H. (ed.) (1931)   *The new survey of London life and labour: London industries I*. London: PS King (volume two).

Llewellyn Smith, H. (ed.) (1932)   *The new survey of London life and labour: survey of social conditions – the eastern area*. London: PS King (volume three).

Llewellyn Smith, H. (ed.) (1933)   *The new survey of London life and labour: London industries II*. London: PS King (volume five).

Llewellyn Smith, H. (ed.) (1934a)   *The new survey of London life and labour: forty years of change*. London: PS King (volume one).

Llewellyn Smith, H. (ed.) (1934b)   *The new survey of London life and labour: survey of social conditions – the western area*. London: PS King (volume six).

Llewellyn Smith, H. (ed.) (1934c)   *The new survey of London life and labour: London industries III*. London: PS King (volume eight).

Lockwood, D. (1986)   Class, status and gender. In Crompton, R. and Mann, M. (eds.) *Gender and stratification*. (Cambridge: Polity), 11–22.

London Amenity and Transport Association (1984)   *The company car factor*. London.

London Amenity and Transport Association and London Motorway Action Group (1971)   'The volume of passenger movement in London'. London: Greater London Development Plan Inquiry Support Document S12/163.

London Borough of Lambeth (1977)   *Waterloo district plan*. London.

London Borough of Lambeth (1984)   *Lambeth local plan*. London.

London Borough of Hammersmith and Fulham and London Borough of Lambeth (1986)   *Sainsburys at Nine Elms: a study of an inner city superstore*. London:

London Borough of Hammersmith and Fulham Research Directorate of Development Planning Report 74.

London Borough of Hillingdon (1984) *Retail warehouse parks.* London: London Borough of Hillingdon Report and Supplementary Guidance Note S1.

London Borough of Hillingdon (1989) *Traffic generation study: high tech estates and office buildings.* London: London Borough of Hillingdon TRICS Report 89/1.

London Chamber of Commerce (1989) *London economy research programme: employment in finance and business services.* London.

London Planning Advisory Committee (1987) *Retailing: report by the topic working party for consultation and discussion.* London.

London Planning Advisory Committee (1987a) *Employment: report by the topic working party.* London.

London Planning Advisory Committee (1988) *Strategic planning advice for London: policies for the 1990s.* London.

London Regional Transport (1989a) *Count of central area peak traffic, Autumn 1988.* London: London Regional Transport Research Memorandum 415.

London Regional Transport (1989b) *Annual report and accounts 1988–89.* London.

London Research Centre (1986) *Review of London's needs.* London: London Borough Grants Committee.

London Research Centre (1988) *London under pressure: a review of needs, provision and voluntary activity.* London: London Borough Grants Committee.

London Tourist Board (1987) *London tourism strategy.* London: London Tourist Board and Convention Bureau.

London Transport (1970) *Annual report 1970.* London: London Transport Executive.

London Transport Executive (1956) *London travel survey 1954.* London.

Low, S.J. (1891) The rise of the suburbs. *Contemporary Review* 40, 545–56.

McAuley, I. (1987) *Guide to ethnic London.* London: Michael Haig.

McDowell, L. and Massey, D. (1984) A woman's place. In Massey, D. and Allen, J. (eds.) *Geography matters!* (Milton Keynes: Open University Press), 128–47.

Mack, J. and Humphries, S. (1985) *London at war.* London: Sidgwick and Jackson.

Mackenzie, S. and Rose, D. (1983) Industrial change, the domestic economy and homelife. In Anderson, J., Duncan, S.S. and Hudson, R. (eds.) *Redundant spaces in cities and regions* (London: Academic Press), 155–200.

Mackinder, H. (1902) *Britain and the British seas.* London: Heinemann.

Mackintosh, M. and Wainwright, H. (eds.) (1987) *A taste of power: the politics of local economics.* London: Verso.

Manley, L. (ed.)(1986) *London in the age of Shakespeare.* London: Croom Helm.

Manners, G. and Morris, D. (1986) *Office policy in Britain: a review.* Norwich: Geo Books.

Mark-Lawson, J. *et al.* (1984) Gender and local politics: struggles over welfare policies 1918–39. In Lancaster Regionalism Group *Localities, class and gender* (London: Pion) 195–215.

Marshall, J.N. (1988) *Services and uneven development.* Oxford: Oxford University Press.

Marshall, M. (1987) *Long waves of regional development.* London: Macmillan.

Martin, J. (1966)   *Greater London: an industrial geography*. London: Bell.

Martin, J. and Roberts, C. (1984)   *Women and employment: a lifetime perspective*. London: HMSO.

Massey, D. (1984)   *Spatial divisions of labour: social structure and the geography of production*. Basingstoke: Macmillan.

Mayhew, J. (1981)   *The Morning Chronicle letters*. Firle: Caliban (volume two).

Ministry of Housing and Local Government (1961)   *London government proposals for reorganisation*. London: HMSO (cmnd 1562).

Morgan, B.S. (1987)   *Shopping centres in the London Borough of Bromley*. London: London Borough of Bromley.

Morgan, W.T.W. (1961)   The two office districts of central London. *Journal of the Town Planning Institute*, 47, 161–66.

Morris, J. (1986)   *Women workers and the sweated trades*. Aldershot: Gower.

Morris, L. (1989)   Employment, the household and social networks. In Gaillie, D. (ed.) *Employment in Britain* (Oxford: Blackwell), 376–405.

Nabarro, R. (1989)   Investment in commercial and industrial development: some recent trends. In Cross, D. and Whitehead, C. (eds.) *Planning and development 1989* (Cambridge: Policy Journals), 65–70.

Nairn, I. (1988)   *Nairn's London: the classic guide revisited by Peter Gasson*. Harmondsworth: Penguin.

Nationwide Anglia Building Society (1989)   *Local housing statistics 1989: report 11, London boroughs*. Swindon.

Nelson, R.L. (1958)   *The selection of retail locations*. New York: Dodge.

Newman, D. (1985)   Integration and ethnic spatial concentrations: changing spatial distribution of the Anglo-Jewish community. *Transactions of the Institute of British Geographers* 10, 360–76.

North, D.C. (1955)   Location theory and regional economic growth. *Journal of Political Economy* 63, 243–58.

Norpoth, H. (1987)   Guns and butter and government popularity in Britain. *American Political Science Review* 81, 949–959.

Norton, A. (1983)   *The government and administration of metropolitan areas in western democracies*. Birmingham: University of Birmingham Institute of Local Government Studies.

Offer, A. (1981)   *Property and politics 1870–1914*. Cambridge: Cambridge University Press.

Office of Population Censuses and Surveys (1981)   *Census 1981: definitions Great Britain*. London: HMSO.

Office of Population Censuses and Surveys (1982)   *1981 Census of Great Britain: Greater London* (CEN81 CR17) London: HMSO.

Office of Population Censuses and Surveys (1983)   *1981 Census of Great Britain: national report part one* (CEN81 NR (1)) London: HMSO.

Office of Population Censuses and Surveys (1989a)   *Key population and vital statistics 1987*. London: HMSO.

Office of Population Censuses and Surveys (1989b)   *England and Wales birth statistics 1987*. London: HMSO.

Office of Population Censuses and Surveys (1990)   *Key population and vital statistics 1988*. London: HMSO.

Offord, J. (1987)   'The fiscal implications of differential population changes for local authorities in England and Wales'. Cambridge: PhD thesis, University of Cambridge.

O'Leary, B. (1987a)   Why was the GLC abolished? *International Journal of Urban and Regional Research* 11, 193–217.

O'Leary, B. (1987b) British farce, French drama and tales of two cities: explaining the reorganization of Paris and London governments 1957–86. *Political Studies* 65, 369–89.

Organization for Economic Cooperation and Development (1983) *Managing urban change: volume one, policies and finance*. Paris.

Organization for Economic Cooperation and Development (1987) *Managing and financing urban services*. Paris.

Orwell, G. (1982) *The lion and the unicorn*. Harmondsworth: Penguin.

Pahl, R.E. (1984) *Divisions of labour*. Oxford: Blackwell.

Parssinen, T. and Prothero, I. (1977) The London tailors' strike of 1834 and the collapse of the Grand National Consolidated Trades Union: a police spy's report. *International Review of Social History* 22, 65–107.

Patterson, S. (1964) Polish London: In Centre for Urban Studies *London: aspects of change* (London: MacGibbon and Kee), 309–43.

Patterson, S. (1965) *Dark strangers: a study of West Indians in London*. Harmondsworth: Penguin.

Peach, C. (1968) *West Indian migration to Britain: a social geography*. Oxford: Oxford University Press.

Peach, C. (1984) The force of West Indian identity in Britain. In Clarke, C., Ley, D. and Peach, C. (eds.) *Geography and ethnic pluralism* (London: Allen and Unwin), 214–30.

Pederson, P.O. (1970) Innovation diffusion within and between national urban systems. *Geographical Analysis* 2, 203–54.

Pharoah, T.M. (1986) The motor car: idol or idle? – a survey of redundancy in domestic car fleets. *Traffic Engineering and Control* 27(2), 64–6.

Pharoah, T.M. and Russell, J. (1989) 'Traffic calming: policy and evaluations in three European countries'. London: South Bank Polytechnic Department of Town Planning Occasional Paper 2/89.

Phillips, D. (1987) The rhetoric of anti-racism in public housing allocation. In Jackson, P. (ed.) *Race and Racism* (London: Allen and Unwin), 212–37.

Pilkington, A. (1984) *Race relations in Britain*. London: University Tutorial Press.

Pirie, M. (1988) *Privatisation: theory, practice and choice*. Aldershot: Wildwood House.

Plender, J. (1982) *That's the way the money goes*. London: André Deutsch.

Plowden, S.P.C. (1980) *Taming traffic*. London: André Deutsch.

Pollard, S. (1983) *The development of the British economy 1914–80*. London: Edward Arnold.

Porter, M. (1990) *The comparative advantage of nations*. London: Macmillan.

Price, D.G. and Blair, A.M. (1989) *The changing geography of the service sector*. London: Belhaven.

Property Market Analysis (1988) *Greater London area retail expenditure and impact study: phase IIA retail market assessment*. London: London Planning Advisory Committee.

Purcell, K. (1989) Gender and the experience of employment. In Gaillie, D. (ed.) *Employment in Britain* (Oxford: Blackwell), 157–86.

Quennell, P. (ed.) (1983) *London's underworld*. London: Bracken Books (original 1862).

Rajan, A. (1988) *Create or abdicate: the City's human resource choice for the 90's*. London: Witherby.

Reiber, B. (1988) *The serious shopper's guide to London*. New York: Prentice-Hall.

Rendel Planning (1988) *Building confidence in London's inner city – a report for the Confederation of British Industry London task force.* London.

Rhodes, G. (1970) *The government of London: the struggle for reform.* London: Weidenfeld and Nicolson.

Robson, B.T. (1973) *Urban growth: an approach.* London: Methuen.

Robson, B.T. (1986) Coming full circle: London versus the rest 1890–1960. In Gordon, G. (ed.) *Regional cities in the UK 1890–1980* (London: Harper and Row), 217–32.

Rogers, B. (1988) *Men only: an investigation of men's organizations.* London: Pandora.

Royal Commission on Local Government in England (1969) *Report.* London: HMSO (cmnd 4040).

Rubinstein, W. (1977) Wealth, elites and the class structure of modern Britain. *Past and Present* 77, 99–126.

Runnymede Trust (1987) *Race and immigration.* London.

Salt, J. (1986) Population trends. In Clout, H.D. and Wood, P.A. (eds.) *London: problems of change* (London: Longman), 52–9.

Samuel, R. (1977) Workshop of the world: steam power and hand technology in mid-Victorian Britain. *History Workshop Journal* 3, 6–72.

Saunders, A. (1988) *The art and architecture of London.* Oxford: Phaidon (second edition).

Saunders, P. (1979) *Urban politics: a sociological interpretation.* Harmondsworth: Penguin.

Savage, M. and Fielding, T. (1989) Class formation and regional development: the 'service class' in South East England. *Geoforum* 20, 203–18.

Schiller, R. (1986) Retail decentralisation – the coming of the third wave. *The Planner* 72, 13–5.

Schiller, R. and Jarrett, A. (1985) A ranking of shopping centres using multiple branch numbers. *Land Development Studies* 2, 53–100.

Schiller, R. and Lambert, S. (1977) The quality of major shopping developments in Britain since 1965. *Estates Gazettes* 242, 359–63.

Schmeichen, J.A. (1984) *Sweated industries and sweated labour: the London clothing trades 1860–1914.* London: Croom Helm.

Scott, A.M. (1986) Industrialisation, gender segregation and stratification theory. In Crompton, R. and Mann, M. (eds.) *Gender and stratification* (Cambridge: Polity), 154–89.

Segrave, C. (1989) *London property guide for buyers, sellers and owners of homes in London.* London: Mitchell Beazley International.

Sekyi, R. (1987) 'Discrimination, real or imagined?: the role of the estate agent in the housing market'. BSc dissertation, King's College London.

Self, P. (1972) Planning. In Rhodes, G. (ed.) *The new government of London – the first five years.* London: Weidenfeld and Nicolson, 299–346.

SERPLAN (1987) *Regional shopping centres around London: background papers: report of the retail monitoring working party.* London.

SERPLAN (1988) *Regional shopping provision in the M25 corridor.* London.

SERPLAN (1989a) *Regional trends in the South East: the South East regional monitor 1988–89.* London: London and South East Regional Planning Conference Report RPC 1430.

SERPLAN (1989b) *Survey of major retail developments completed, committed or proposed since 1980* (technical appendix to SERPLAN 1989a). London.

SERPLAN (1989c) *Town centre quality survey 1988: set of tables* (technical appendix to SERPLAN 1989a). London: London and South East Regional Planning Conference Report RPC 1478.

SERPLAN (1989d)   *Town centre quality survey 1988: detailed findings.* London: London and South East Regional Planning Conference Report RPC 1442.

Sharpe, L.J. (1965)   *Research in local government.* London: London School of Economics Greater London Paper 10

Shepherd, J. *et al.* (1974)   *A social atlas of London.* Oxford: Oxford University Press.

Simmons, I. (1981)   Contrasts in Asian residential segregation. In Jackson, P. and Smith, S.J. (eds.) *Social interaction and ethnic segregation* (London: Institute of British Geographers Special Publication 12), 81–100.

Smailes, A.E. and Hartley, G. (1961)   Shopping centres in the Greater London area. *Transactions of the Institute of British Geographers* 29, 201–13.

Smart, M.W. (1974)   Labour market areas: uses and definition. *Progress in Planning* 2, 239–353.

Smith, D.H. (1933)   *The industries of Greater London.* London: PS King and Sons.

Smith, G. (1986)   *Retail warehouses in London.* London: London Research Centre Reviews and Studies Series 30.

Smith, N. and Williams, P. (eds.) (1986)   *Gentrification of the city.* London: Allen and Unwin.

Smith, S.J. (1987)   A geography of English racism. In Jackson, P. (ed.) *Race and racism* (London: Academic), 25–49.

Southall, H. (1988)   The origins of the depressed areas: unemployment, growth and regional economic structure in Britain before 1914. *Economic History Review* 41, 236–58.

Spate, O.H.K. (1938)   Geographical aspects of the industrial evolution of London since 1850. *Geographical Journal* 92, 422–32.

Stocks, M. (1939)   London government. *Political Quarterly* 10, 365–374.

Stopford, J. and Dunning, J.H. (1958)   *Multinationals: company performance and global trends.* London: Macmillan.

Stow, J. (1958)   *The survey of London.* London: Dent (1956 edition)

Sumner, J. and Davies, K. (1978)   Hypermarkets and superstores: what do the planning authorities really think? *Retail and Distribution Management* 6, 8–15.

Sutcliffe, D. (1982)   *British black English.* Oxford: Oxford University Press.

Talbot, J. (1988)   Enterprise zones – are there no lessons for inner city policy? *The Planner* 74(2), 64–7.

Taylor, P.J. (1989)   Britain's century of decline: a world-systems interpretation. In Anderson, J. and Cochrane, A. (eds.) *A state of crisis: the changing face of British politics* (London: Hodder and Stoughton), 8–26.

Thane, P. (1981)   Social history 1860–1914. In Floud, R. and McCloskey, D. (eds.) *The economic history of Britain since 1700.* (Cambridge: Cambridge University Press), 198–238 (volume two).

Thom, D. (1989)   Free from chains? The image of women's labour in London 1900–20. In Feldman, D. and Jones, G.S. (eds.)*Metropolis. London* (London: Routledge), 85–99.

Thomson, J.M. (1969)   *Motorways in London.* London: Duckworth.

Thomson, J.M. (1977)   *Great cities and their traffic.* London: Gollancz.

Thorpe, D. and Kivell, P.T. (1974)   'Atlas of Greater London shopping centres'. Manchester: Manchester Business School Retail Outlets Research Unit Research Report 7.

Townsend, P. *et al.* (1988)   *Health and deprivation: inequality and the North.* London: Croom Helm.

Trench, R. and Hillman, E. (1984)   *London under London: a subterranean guide.*
London: John Murray.

Tucker, R.S. (1936)   Real wages of artisans in London 1729–1935. *Journal of the
American Statistical Association* 31, 73–84.

Tunley, P. *et al.* (1979)   *Depriving the deprived: a study of finance, educational
provision and deprivation in a London borough.* London: Kogan Page.

Turner, J.E. (1978)   *Labour's doorstep politics in London.* London: Macmillan.

Unit for Retail Planning Information (1988)   *List of UK hypermarkets and
superstores.* Reading.

Unit for Retail Planning Information (1989)   miscellaneous promotional literature for The Data Consultancy dated October 1989. Reading.

United Nations Industrial Development Organization (1983)   *Industry in a
changing world.* New York.

Valpy, R.A. (1889)   *An inquiry into the conditions and occupations of the people
in central London.* London: Stanford.

Vance, J.E. (1962)   Emerging patterns of commercial structure in American
cities. In K. Norborg (ed.) *Proceedings of the International Geographical
Union symposium on urban geography: Lund 1960* (Lund: C.W.K. Gleerup),
485–518.

Vercoe, E. (1988)   *Where to live in London 1988: the first independent guide to
choosing a London home.* London: Sidgwick and Jackson.

*Walby, S. (1987)   Patriarchy at work: patriarchal and capitalist relations in
employment.* Cambridge: Polity.

Walby, S. (1989)   Theorising patriarchy. *Sociology* 23, 213–34.

Walker, D. (1983a)   *Municipal empire: the town halls and their beneficiaries.*
London: Temple Smith.

Walker, D. (1983b)   Local interest and representation: the case of 'class' interest
among Labour representatives in inner London. *Environment and Planning
C: Government and Policy* 1, 341–60.

Walker, R. (1978)   The transformation of urban structure in the nineteenth
century and the beginning of suburbanization. In Cox, K.R. (ed.) *Urbanization
and conflict in market societies* (London: Methuen), 165–212.

Ward, S.V. (1988)   *The geography of inter-war Britain.* London: Routledge and
Kegan Paul.

Warnes, A.M. (1980)   The long-term view of employment decentralisation from
the larger English cities. In Evans, A. and Eversley, D.E.C. (eds.) *The inner
city* (London: Heinemann) 25–44.

Warnes, A.M. (1987)   The distribution of the elderly in Great Britain. *Espace
Populations Sociétés* 1987/1, 197–212.

Warnes, A.M. (1989)   Social problems of elderly people in cities. In Herbert,
D.T. and Smith, D.M. (eds.) *Social problems in the city* (Oxford: Oxford
University Press), 197–212 (second edition).

Waterman, S. and Kosmin, B. (1987)   Ethnic identity, residential concentration
and social welfare. In Jackson, P. (ed.) *Race and racism* (London: Academic),
254–71.

Weber, A.F. (1899)   *The growth of cities in the nineteenth century.* New York:
Macmillan.

Weightman, G. and Humphries, S. (1984)   *The making of modern London
1914–39.* London: Sidgwick and Jackson.

Wells, H.G. (1901)   *Anticipations of the reactions of mechanical and scientific
progress upon human life and thought.* London: Chapman and Hall.

Westaway, J. (1974)   Contact potential and the occupational structure of the
British urban system. *Regional Studies* 8(2), 57–73.

Whitt, J.A. (1987)   Mozart in the metropolis: the arts coalition and the urban growth machine. *Urban Affairs Quarterly* 23, 15–36.

Wilson, J. (1984)   Retail warehousing in Greater London. *Estates Gazette* 272, 244–46.

Wheen, F. (1985)   *The battle for London*. London: Pluto Press.

Witherspoon, S. (1985)   Sex roles and gender issues. In Jowell, R., Witherspoon, S. and Brook, L. (eds.) *British social attitudes: 1985 report* (Aldershot: Gower), 55–94.

Witherspoon, S. (1988)   A woman's work. In Jowell, R., Witherspoon, S. and Brook, L. (eds.) *British social attitudes: 1988 report* (Aldershot: Gower), 175–200.

Wohl, A.S. (1977)   *The eternal slum: housing and social policy in Victorian London*. London: Edward Arnold.

Wood, P.A. (1986)   Economic change. In Clout, H.D. and Wood, P.A. (eds.) *London: problems of change* (London: Longman), 60–74.

World Bank (1988)   *The world development report 1988*. Washington D.C.

Wrigley, E.A. (1967)   A simple model of London's importance in changing English society and economy 1650–1750. *Past and Present* 37, 44–70.

Yelling, J. (1990)   The metropolitan slum: London 1918–51. In Gaskell, M. (ed.) *The slum* (Leicester: Leicester University Press).

Young, K. (1975a)   The Conservative strategy for London, 1855–1975. *London Journal* 1(1), 56–81.

Young, K. (1975b)   *Local politics and the rise of the party: the London Municipal Society and the Conservative intervention in local elections 1894–1963.* Leicester: Leicester University Press.

Young, K. (1980)   Implementing an urban strategy: the case of public housing in metropolitan London. In Ashford, D. (ed.) *National resources and urban policy* (New York: Methuen), 239–57.

Young, K. and Garside, P.L. (1982)   *Metropolitan London: political and urban change 1837–1981*. London: Edward Arnold.

Young, K. and Kramer, J. (1978)   Local exclusionary politics in Britain: the case of suburban defence in a metropolitan system. In Cox, K.R. (ed.) *Urbanization and conflict in market societies* (London: Methuen), 229–51.

Young, M. and Willmott, P. (1964)   *Family and kinship in East London*. Harmondsworth: Penguin (revised edition).

Young, T. (1934)   *Becontree and Dagenham: the story of the growth of a housing estate*. London: Pilgrim Trust.

Zeff, L. (1986)   *Jewish London*. London: Piatkus.

Zweig, F. (1952)   *The British Worker*. Harmondsworth: Penguin.

Index